The Builder's Pocket Manual

THE

BUILDER'S POCKET MANUAL;

CONTAINING THE ELEMENTS OF

BUILDING, SURVEYING,

AND

ARCHITECTURE;

WITH

PRACTICAL RULES AND INSTRUCTIONS

IN

CARPENTRY, BRICKLAYING, MASONRY, &c.

OBSERVATIONS ON THE PROPERTIES OF MATERIALS,

AND

A VARIETY OF USEFUL TABLES AND RECEIPTS.

WITH TWELVE PLATES.

By A. C. SMEATON,

CIVIL ENGINEER, &C.

A NEW EDITION, REVISED, CORRECTED, AND ENLARGED.

LONDON:

M. TAYLOR, BARNARD'S INN, HOLBORN.
NEW YORK: WILLIAM JACKSON, CEDAR STREET.

MDCCCXXXVII.

WILLIAM JACKSON, PRINTER.

PREFACE.

ALTHOUGH so many books have been written upon Architecture and the Art of Building, calculated to assist the experienced practitioner, there are few, if any, that profess to assist the student in the acquisition of elementary knowledge. This important omission is supplied by this volume, and it will be found, we hope, a desirable assistant to every young man who has devoted himself to any engagement connected with these interesting pursuits. If the reader desire to become a workman, he will find, in these pages, facts that will aid his progress, and convey in a few hours the information that has been collected with great labour by studious men, in a long series of years; if he intend to be a surveyor of work, he will here find the elements of his art. The architectural student will need both these branches of knowledge, but to assist him still more, we have added an outline of the history of his profession. To whatever branch of the art of building the reader may belong, he will find something valuable to him, and calculated to assist his progress. But, although the book is mainly intended for the student, yet it is hoped that every workman who peruses it will gather some valuable information, and some practical hints which he may carry out in his engagements. To the amateur also, the following pages may be useful, by explaining the technical terms used among workmen, as well as the scientific principles which regulate construction.

ADVERTISEMENT TO THE SECOND AMERICAN EDITION.

THE extremely rapid sale of the former edition of this work having convinced the Publisher that it was adapted in a considerable degree to the wants of the Artificers for whom it was chiefly compiled, he has been induced to have the whole revised, in order to present it again to the Public in as complete a form as circumstances would permit. It has therefore been placed in the hands of one whom he considered competent to the task, and the results of whose labors will be found in the ensuing pages.

It is too commonly the case, in producing works of this nature, to consider the *matter* only, without regard to the *manner*, of the information communicated. Hence the expression is often vague, the style hard, and the subject is robbed of the interest to which its importance is entitled ; for, be it remembered, the greater proportion of those who read works of this kind are not such as will prosecute their search of information through all the difficulties of language. An attempt has been made to remedy the defects of the former edition, in this respect, not by any alteration in the text, such as the substitution of new matter for old, but by occasionally altering the term of the expression, and by additional care with regard to the punctuation.

The work being originally written for English perusal and practice, was deficient in some respects for the necessities of this country ; to supply such deficiencies, and to adapt it more closely to American use, a few notes have been appended to the text, brief accounts of certain materials which are used here, but not in Great Britain, have been added, and rigid care has been taken as to the typography. The calculations also have been revised and corrected. In England, where this work originally appeared, it could hardly have any other object than that of being an assistant to the workman, as it is usual there to call in the surveyor, before payment is made for work of any magnitude. But in the wide extent of this country, where artificers are comparatively scarce, and in particular the western and all new regions, it is incumbent on the settler to put his own hand to many things of which in a crowded population he never dreamt. Hence a knowledge of the details, in matters so important as those which are here contained, will be found of the greatest advantage. They will facilitate the acquisition of those domestic comforts without which the arduous duties of clearing and settlement could not be successfully encountered ; they will enable him also to do this cheaply, effectually, and quickly, all of which are highly important considerations.

It is therefore confidently believed that the present edition of The Builder's Manual will be found considerably improved and generally useful. It is offered in the expectation that artificers will meet with occasional hints and information on subjects connected with their particular occupations, and that the community will find their judgment and calculations materially assisted when purposing to incur expenses of such nature.

CONTENTS.

INTRODUCTION.

So intimately is the art of building connected with a provision for the comforts and conveniences of life, that it has engaged the attention of men from the period when they first formed themselves into societies. In the early ages of the world, little more could have been required than a temporary shelter from occasional atmospheric changes, and houses or huts were probably constructed in a very rude and imperfect manner; but as even communities were not then accustomed to confine themselves to any locality, such residences were sufficient for their purposes. But when large societies determined to occupy a place as a constant residence, they surrounded themselves with all those permanent comforts which might be within their reach. The art of building necessarily attracted much of their attention, and nations vied with each other in an attempt to blend stability of structure with elegance of appearance. These are the objects of builders in the present day; but at the same time, the altered state of society requires that they should be equally careful to

B

secure economy in the use of materials, that no unnecessary expense may be incurred by their waste or
misapplication, or by the addition of unnecessary labour.

The importance of the subject has induced men to
acquaint themselves with the general principles of construction, and the application of ornament; and to give
their attention to individual branches of the science and
art of building, so as to obtain, by the combined labours
of many, some knowledge of the whole. Many expensive and useful books have been published, by both
architects and builders, upon different subjects connected with the art and science of building; but many
of these books are not only too costly for the means
of some persons desirous of knowledge, but would be
almost useless if they could be obtained. A prelimi-
nary knowledge is required before the student can either
perceive the importance of the information they contain,
or the means by which it may be applied. There are,
it is true, many introductory books, but they chiefly
treat of Architecture and Designing, and are of little
assistance to the workman or the student.

In preparing this manual, the author has endeavoured
to supply the reader with such important elementary
knowledge, as shall enable him to understand the general principles of building, and fit him for the perusal of
those works which have been written on the several
subjects connected with the art. There are three
classes of men engaged in the completion of a build-

ing, the architect, the builder, and the surveyor; and
each should be perfectly acquainted with the business
of the others. Some persons have professed the three
arts, a practice which cannot be too strongly condemn-
ed, since it is impossible that any man can give sufficient attention to all, to do either, correctly or well.
But at the same time an acquaintance with all is
desirable, for they are so closely connected, that one
cannot be properly practised without the assistance of
the others.

The business of the Architect is to design buildings,
to make such drawings, and to so describe them, as
shall enable the builder to execute that which he has
planned. The surveyor measures the work when
finished, and affixes appropriate prices, according to his
judgment of the manner in which the workman has performed his task, and the difficulties which have attended
the execution. An elementary work on Building should
describe the manner in which these persons severally
perform their tasks, and we have therefore divided our
book into three parts or sections, which we have designated the Builder, the Surveyor, and the Architect.

THE

BUILDER'S MANUAL.

THE BUILDER.

WRITERS on Architecture have frequently divided the art into three parts, because in the erection of a building three things are required, *strength*, *convenience*, and *beauty*.

In order to obtain strength, good materials must be employed, and they must be well applied. There must be a proper arrangement of the several portions of edifices, so that instead of weighing down or oppressing each other, they may mutually strengthen each other; and should faults be suspected to exist, in either the quality or dimensions of the materials used, they must be employed where they would be sufficient for the purpose, should the suspicion be realized. The builder must also be careful that any stress may be met by a suitable arrangement of parts, and that the strength may be in a reciprocal proportion to the stress which is to be overcome.

To provide convenience, the building must be suited to the purpose for which it is intended. The rooms, for instance, should be of a size proportionate to the use for which they are to be employed, or the business that is to be done in them; a small house should not be encumbered and lessened with a large staircase, nor a large mansion be rendered uncomfortable by one that is cramped in its dimensions. " The hall," as Fuller says, " ought to lie open, and so ought galleries and stairs, provided the whole house be not spent in paths. Chambers and closets ought to be private and retired." Every part should be suited to the purpose for which it is to be used.

The beauty of a building does not altogether depend upon its architectural decorations and ornaments; but there must be a just proportion of all its parts, the width, length, and height, being everywhere so adjusted, as to produce that harmony calculated to give pleasure to the observer. Many persons err in overloading an edifice with ornament, while others impair the general appearance by neglecting altogether its enrichment. There should never be introduced an ornament, that has the appearance of supporting a weight, where there is evidently no weight to support; and when mouldings are employed, they should have an agreement with the dimensions of the walls on which they are to be fixed, being neither heavy in small apartments, nor diminutive in large ones.

The first thing to be done when a building is to be erected, is to survey the ground on which it is to be placed, with a view to determine the nature of the soil, whether it be rocky, swampy, or composed of clay, gravel, or sand. When this has been determined, the foundations may be arranged for, and the operations required must be regulated accordingly.

The dimensions must then be set out, as shown upon the plan of the basement. This is best done by first marking out the line of the principal front, and then placing stumps, or pins, at those parts where the side and internal walls meet it. When the several angles have been determined, and the line of walls marked out, the excavator may proceed to form the trenches which are to receive the footings, or foundations; and the work is then regularly proceeded with, according to the drawings which are placed in the workman's hands. And here it may be necessary to remark, that architects generally form their drawings from a scale of one eighth of an inch to a foot; but this is not adopted in every case; and, therefore, to prevent mistake, the plans and elevations are generally figured. The scale of one inch to a foot is the most convenient for workmen, for they have then only to apply their rule to the several parts of the drawing, and calculating every inch as a foot, it is scarcely possible for them to make a mistake. But it is not always practicable to draw a plan to this scale, as it would in some instances extend the drawing to an inconvenient size.

These general remarks may be of some service to the beginner, as illustrating the objects to be obtained in building, and the manner in which the workman is to commence his operations. We may now proceed to make some more particular remarks upon the several departments of building, the nature and composition of the materials employed in each, and the methods by which they are worked. As this little volume is intended for the use of the student in all departments, we shall not consider any fact, however self-evident it may appear, too simple to be mentioned; but we shall endeavour to lead him on, by easy steps, from the simple to the more complex principles of the art, giving so much of the science as may appear necessary to afford a reason for the process that may be adopted.

THE BRICKLAYER.

As the art of Bricklaying is generally supposed to be so simple as to require little or no attention, it will be necessary to remove this false impression, by a somewhat particular detail of the facts which relate to it. There are many persons, and even some workmen, who suppose that nothing more is required than that the bricks should be properly bedded, and the work level and perpendicular. But the workman who would attain perfection in his business, should acquaint himself with the different arrangements made use of in placing the bricks, so that one part of the work shall strengthen another, and thus prevent one portion from a greater liability to give way than another. It is also necessary that the workman should be acquainted with the several sorts of bricks, their qualities, and the uses for which they are particularly adapted.

It appears from history, that bricks have been employed for building, from a very early period. We are informed by the sacred records, that, very shortly after the occurrence of that universal catastrophe, which swept from the earth nearly the whole human race, and remodelled its surface, the sons of Noah fixed their abode in a plain in the land of Shinar or Chaldea; " and they said one to another, go to, let us make brick, and burn them thoroughly. And they had brick for stone, and lime had they for mortar." By the same authority we are informed that the Jews, during their servitude to the Egyptians, were employed not only in making bricks, but also in building with them. " And they (the Egyptians) made their lives bitter with hard bondage, in mortar, and in brick."—"And they built for Pharoah treasure cities, Pithom and Raamses." Nearly all the Egyptian buildings, spared by the devastating hand of time, are constructed of stone, but there are some brick buildings still in existence, and Pocock mentions a pyramid constructed of unburnt brick.

From all the evidence we can collect on the subject, except that to which we have referred, it does not appear that the Egyptians, or any other of the early inhabitants of the earth, were acquainted with the art of burning bricks; but both the Greeks and the Romans used them. Vitruvius has given a description of the kind of bricks used in his own day, and has offered some suggestions as to the choice of the material from which they ought to be formed. The passage is interesting, and as the works of this author may not be in the possession of all our readers, we may be permitted to quote it from Mr. Gwilt's translation. "They should be made of earth, of a red or white chalky, or a strong sandy, nature. These sorts of earth are ductile and cohesive, and not being heavy, bricks made of them are more easily handled in carrying up the work. The proper seasons for brick-making are the spring and autumn, because they then dry more equally. Those made in the summer solstice are defective, because the heat of the sun soon imparts to their external surfaces an appearance of sufficient dryness, whilst the internal parts of them are in a very different state; hence when thoroughly dry, they shrink and break those parts which were dry in the first instance; and thus broken, their strength is gone. When plastering is laid and set hard on bricks which are not perfectly dry, the bricks which will naturally shrink, and consequently occupy a less space than the plastering, will thus leave the latter to stand of itself. It is not therefore without reason that the inhabitants of Utica allow no bricks to be used in their buildings, which are not at least five years old, and also approved by a magistrate.

"There are three sorts of bricks; the first is that which the Greeks call Didoron, being the sort we use, that is, one foot long, and half a foot wide. The two other sorts are used in Grecian buildings; one is called Pentadoron, the other Tetradoron. By the word Doron, the Greeks mean a palm. That sort which is five palms each way is called Pentadoron; that of four palms, Tetradoron. The former of these two sorts are used in public buildings, the latter in private. Each sort has half bricks made to suit it, so that when a wall is executed, the course on one of the faces of the wall shows sides of whole bricks, the other face of half bricks, and being worked to the line on each face, the bricks on each bed bend alternately over the course below."

There has been some dispute among antiquaries as to the time when bricks were first introduced into England. Dr. Lyttleton states, in the Archæologia, that there were no brick buildings earlier than the fourteenth century. Bagford says they were introduced in the reign of Henry the Seventh, but it must have been earlier than this, for Ewelme palace in Oxfordshire, erected by William de la Pole, and Herstmonceaux Castle in Sussex, were both erected in the reign of Henry the Sixth. But we leave the antiquaries to determine this disputed question, and proceed to make a few remarks of a more practical character.

Bricks.

Brick is an artificial stone, formed of clay, moulded in rectangular prisms of constant dimensions, and hardened by burning, or exposure to the sun. All bricks made in England must be, according to act of parliament, nine inches long, four and a half inches broad, and two and a half thick.

There are several kinds of bricks; the most important to be mentioned are marls, stocks, and plàce bricks. All these are formed in moulds of the same size, and differ only in quality, which depends upon the character of the clay, the care taken in tempering it, and the manner in which it is burnt. The best marls are called firsts, and are used for the heads of doors and windows; the seconds are used for facing, that is, for the front of a building; and for this purpose they are admirably adapted, not only on account of their colour, which is a yellowish white, but also for their compactness, and capability of resisting the action of the atmosphere. Grey stocks are sometimes used instead of marls, but they are of inferior quality. Place bricks are the refuse of a burning, and are in fact those which have not been perfectly burnt. Clinkers are overburnt bricks. For paving, Dutch clinkers, so called because imported from Holland, are frequently used; they are very hard, and have a light yellow colour. These bricks are six inches long, three inches broad, and are laid herring-bone way.

Tiles.

There are several sorts of tiles. Paving-tiles, used for kitchens and dairies in farmhouses, are about nine inches long, four and a half broad, and one and a half thick. Roofing-tiles are formed in different ways, and are known as pan-tiles, plain-tiles, hip-tiles, and ridge-tiles.

Pan, or Flemish-tiles, are fourteen inches and a half long, and ten and a half broad. It is seldom that these tiles are used, even in country towns, for any other purpose than that of covering sheds and out-houses; and, as they have no pin-holes, they are altogether unfit for a high-pitched roof.

The size of plain-tiles is regulated by law, and they should be ten inches and a half long, six and a quarter broad, and five eighths of an inch thick. They are hung on the laths by oak pins, there being two holes in each tile.

Ridge and hip-tiles are of a semi-cylindrical form, and are thirteen inches long, and sixteen inches girt on the exterior surface.

Brick-making.

Bricks should be made of an earthy loam; but the manufacturer is not generally very careful as to the earth he uses, so that it be only possible to make an article which he can sell, or employ himself. Hence it is that some bricks are very brittle, because there is too large a quantity of sand; and others are shaky, because they contain too little, and crack in the drying. It is absolutely necessary, for the manufacture of a good brick, that the earth of which it is to be formed, should be exposed to the air, and especially to the frosts of winter, at least during one year, that it may be pulverized, as this will aid the tempering; and the more it is turned over, during the time of its exposure, the better will be the brick.

An experiment, made by M. Gallon, fully proves the necessity of well tempering the earth to be employed in brick-making. "He took a certain quantity of the earth, prepared for the making of bricks, he let it remain for seven hours, then caused it to be moistened and beaten during the space of thirty minutes; the next morning the same operation was repeated, and the earth was beaten for thirty minutes; in the afternoon it was beaten for fifteen minutes." After moulding a brick, made of this earth, he found that it weighed five pounds eleven ounces, but one made of the same earth without the same preparation, weighed five pounds seven ounces. When the bricks were dried and burnt, he tested their strength, and found that, under the same circumstances, the brick made of well-tempered clay broke with a weight of one hundred and thirty pounds, while the other broke with a weight of seventy pounds. This result clearly proves the necessity of well tempering the brick earth, which is usually done by a mill, put into motion by horses.

When the clay is prepared, it is pressed into a mould, ten inches in length, and five in breadth ; but the brick itself, when burnt, is not more than nine inches long, and four and a half broad, on account of the contraction it suffers by exposure to heat, driving off the water which is in combination with the clay. When the bricks are turned from the mould, which is readily done, the mould being strewed with sand to prevent the adhesion of the clay, they are placed in hacks in a diagonal position, so as to admit the air. Each hack is two bricks wide, and eight bricks, on edge, high. To prevent the access of rain, long sheds are sometimes erected, and the hacks are formed under them ; but at other times they are covered with wheat or rye straw. The time required to dry the bricks must depend upon the weather ; if favourable, it may be done in six or eight days.

Bricks are burnt either in clamps or kilns ; the former are generally used, but the latter are preferable.

Clamps are made with the bricks to be burnt. The foundation is made with place bricks, and of an oblong form. The flue is first formed, passing through the clamp, and about a brick wide. Above each course of brick, a layer of cinders or breeze is placed, the bricks being

c

placed diagonally about an inch apart on each side of the flue. When the clamp is about six feet high, a second flue is made similar to the other, that is to say, if the bricks are immediately required, if not, the flues may be placed about nine feet apart; each flue being filled with coal, breeze, and wood, closely pressed. A layer of breeze is always laid at the top of the whole. The fire-places are usually placed on the western side of the clamp. The bricks may, if required, be burnt in twenty or thirty days, the time varying according to external circumstances. The outside of the clamp is sometimes plastered with clay when the weather is precarious.

Kilns are frequently used for burning bricks, but more commonly in the country than in the neighbourhood of London. They are to be preferred to clamps, as they require less fuel, and less time is required in the process. The walls of a kiln incline inwards, and are usually a brick and a half thick. A kiln is about thirteen feet long, ten feet wide, and twelve feet high, and will burn about twenty thousand bricks at the same time. The bricks are laid upon an open floor, and after they have been thoroughly dried by a gentle fire, a pile of bricks, closed with wet earth, is placed before the fireplace, space being left to add faggots as may be required. When the arches have a white heat, and fire appears at the top, the heat is slackened and then increased, until the bricks are thoroughly burnt, which is generally in about three days. The workman can always determine whether the bricks are dried or not, by the colour of the smoke, which turns from a darkish to a transparent colour, as soon as this has been accomplished; the burning is then commenced.

The advantages which result from a division of labour are well known, and they are not more evident in any mechanical employment, than in the manufacture of bricks. In a long day, that is to say, between five in the morning, and eight at night, a good moulder, will produce five thousand bricks.

There is a very judicious remark in Mr. Partington's Builder's Complete Guide, but we are at a loss to say whether we are indebted to him, or to Mr. Malcolm, for it; we have quoted it as it stands in the work we have named. "The colour of London bricks is not red, as is the case with the common bricks and tiles, but of a light brownish yellow. This colour is more pleasing to the eye than that of the common red brick, and on this account the London bricks are preferred for building houses. The brick-masters assign a curious reason for this colour. According to them, their bricks are kept as much as possible from the contact of the air, during the burning. The consequence of this is, that the iron contained in them is not oxidized to so great a degree as in common bricks; but this mode of reasoning is far from exact. If air were entirely excluded, the bricks would not be burnt at all; because the fire would be extinguished. But if enough air be admitted to burn the coal, mixed with the clay, (which must be the case,) that air must also act upon the iron, and reduce it to the state of a peroxide; indeed there can be no doubt, that the iron in the London yellow bricks is in the state of a peroxide, as well as in the red bricks; for the peroxide of iron gives various colours to bodies, according to circumstances. With it, we find bodies tinged red, yellow, and brown, according to the substances with which the oxide is combined. We ascribe the colour of the London bricks to the ashes of the coals, which, by uniting with the peroxides of iron, form a kind of yellow ochre."

A patent was sometime since taken out by Mr. Shaw for the manufacture of bricks. This gentleman proposed a very ingenious arrangement, by which the clay could not only be pressed into the mould without manual labour, but be also removed by machinery. The machinery may be moved by any mechanical power, whether it be manual, steam, or horse.

CEMENTS.

Having explained the manner in which bricks are made, and the means of distinguishing their qualities, it will be necessary to state the composition of the several kinds of cement, that are used in order to bind or connect the several parts together; and it may here be necessary to mention, that we shall not confine our remarks to those cements which may be used by the bricklayer, but shall also refer to those whieh may be commonly employed by the mason; for as we must speak of the origin of the cementitious principle, it seems desirable to explain all the several kinds of substances, in the composition of which this principle is called into action. But before we speak of the cements themselves, it will be necessary to refer to the nature of that substance, Lime, which is their principal ingredient.

Lime.

Lime is easily distinguished from other substances by its properties. It is an earth having a white colour, and produces a caustic sensation upon the tongue; is incapable of fusion by ordinary temperatures, being one of the most infusible substances in nature, and is but little soluble in water, though it is more soluble in cold than in hot water. Lime is seldom, if ever, found pure in nature, but is generally in combination with an acid; most frequently with carbonic acid, as in the formation of chalk, limestone, and marble. Lime is a very abundant ingredient in the composition of the earth's crust, and generally makes its appearance as a carbonate, but both sulphates and carbonates of lime are found to occur as constituent parts of mineral substances. To obtain pure lime, that is, lime separated from an acid, with which it is uniformly combined in nature, the mineral must be submitted to a red heat, which drives off the acid, and leaves the lime in a state of purity; it is then called caustic or quick-lime. Chalk, lime-stone, marble, oyster-shells, and other substances, are carbonates of lime; and either of these will, when burnt, furnish the material required in building; but the two former are chiefly used for this purpose.

Builders are well aware of the fact, that all limestones or mineral substances containing lime, as an ingredient, do not possess the same cementitious properties. One stone may yield, when burnt, a lime very superior to another, and this difference depends upon the quantity and character of the adventitious substances, which are combined with the lime. Many of these may be detected by the appearance of the mineral, or by very simple experiments. When the limestone has a deep brown or red colour, it generally contains iron, and when burnt has a yellowish hue; when it does not freely effervesce with the application of an acid, and is sufficiently hard to scratch glass, it contains silex; when it effervesces slowly, and gives a milky appearance to the acid, it contains magnesia. The effects of these and other substances upon cements, have not been accurately determined.

The cementing quality of lime seems to arise from its chemical combination with the substances with which it is mixed. First of all it unites with a certain proportion of water, forming a hydrate of lime, which appears to have

c *

a chemical attraction for silica, that is to say, the sand
with which it is mixed.   After exposure to the atmosphere for a short time, it abstracts and applies a portion
of carbonic acid, which greatly increases its hardness,
and on this account all old mortars are remarkable for
their cohesion and strength, frequently becoming stronger
than the stones they unite.   Sir Humphrey Davy, speaking of cements, says, " The cements which act by combining with carbonic acid, or the common mortars, are
made by mixing together slaked lime and sand.   These
mortars at first solidify as hydrates, and are slowly converted into carbonate of lime, by the action of the carbonic acid of the air.   Mr. Tennant found that a mortar
of this kind, in three years and a quarter, had regained
63 per cent of the quantity of carbonic acid which con-
stitutes the definite proportion in carbonate of lime."
But there are two kinds of cement used in building; that
in which lime forms a prominent combination with
water, and this is called a water cement; and that which
combines with carbonic acid, which is called a mortar:
this distinction is a very important one; one kind has
the property of setting under water, the other has not.

Sand.

Sand is a very important ingredient in cements, and too
much pains cannot be taken to obtain it pure.   River
sand should always be preferred to pit sand, for it is less
likely to be mixed with clayey or other substances, which
greatly injure the indurating property of the cement.
But wherever the sand may be obtained, it should be well
washed, and this is especially necessary if taken from the
sea; for the salt with which it is combined, having strong
hygrometric properties, would prevent the cement from

drying. This effect we remember to have frequently observed in a little seaport town, where beach-sand had been used by the builders, without sufficient washing..

Mortar.

Mortar is made of lime and sand, thoroughly mixed together, and brought into the consistency of a paste, by the addition of water. Different proportions of these substances are used by builders, and this must necessarily be the case, for a larger or smaller quantity of sand must be added in proportion to the quality of the lime. A good lime will take more sand than a bad one, and the value of the cement may, in a great measure, be judged of by the quantity of sand it contains. Builders are accustomed, for instance, to use more sand with stone-lime than with chalk-lime; not that there is in general much difference between the two, when first burnt, but because the quality of the chalk-lime is speedily injured by a very rapid absorption of carbonic acid. With one hundred and fifty pecks, that is, thirty-seven and a half striked bushels of chalk-lime, the workman mixes two loads of sand, each load consisting of thirty striked bushels; but twenty bushels of stone-lime will frequently bear two loads and a half of sand. It is estimated that the mortar produced by either of these proportions will do a rod of brickwork, that is, two hundred and seventy-two and a quarter square feet, superficial measure, a brick and half thick, that is, about fourteen inches. According to the experiments of Dr. Higgins, a proportion of one peck of lime to seven of sand, makes the best mortar.

When mortar is to be used in a situation where it will dry quickly, it should be made with as little water as possible, but it is better that the mortar should dry gradually and slowly, as it then becomes more indurated. It is stated by some writers that mortar is injured by keeping, and under one condition, exposure to the air, it is; but, if excluded from the air, it is rather benefited than injured. Pliny states, that the Roman builders were prohibited by law from using a mortar that was less than three years old; and attributes the stability of all their large buildings to this circumstance. But when old mortar is used, it should be well beaten up before it is employed. The reader must not, however, suppose that these remarks justify the exposure of mortar to the air for a considerable time before it is used, a practice very common, but highly improper. This practice probably arose from the difficulty which workmen sometimes find in slaking the lime, in consequence of its being insufficiently burnt, or containing a large portion of argillaceous matter. But above all other things, it is important to use good lime, and to soak the bricks which are to be bedded, before they are laid; for, if the bricks are dry, they imbibe the moisture of the cement, and destroy its quality. There are two things which cause mortar and cements generally to crack, too small a quantity of sand, and a too rapid exhalation of the water. There must always be a contraction, but it is least in those mortars which contain the greatest proportion of sand; for it is the moistened lime which contracts during the process of drying. All mortars may, for a time, be affected by atmospheric changes, and especially by alternate wetting and freezing; but this is most remarkable in those which are liable to crack. A mortar which sets without cracking will always stand afterwards.

Dr. Higgins, to whom we are much indebted for his experiments upon cements, invented one which he speaks of as admirably adapted for both internal and external work, and becomes as hard as Portland stone, when dry. "Take," he says, "fifty-six pounds of coarse sand, and forty-two pounds of fine sand; mix them on a large plank of hard wood, placed horizontally; then spread the sand so that it may stand at the height of six inches, with a flat surface, on the plank; wet it with the cementing liquor; to the wetted sand add fourteen pounds of the purified lime, in several successive portions, mixing and beating them together; then add fourteen pounds of the bone-ash in successive portions, mixing and beating all together." Whatever may be the quality of this cement, it is not likely ever to come into general use, as it would be more expensive, and give more trouble in preparation, than many others which are now found to answer the builder's purpose. This, however, was proposed as a water-cement. Mortar is evidently unfit to be used in any situation where the force of water is to be resisted; for although it is said that mortar composed of lime and sand, in the proportion of one and seven, will not suffer from water, yet, as this composition is seldom, if ever, obtained, it would be folly to risk the security of a building by its use.

The insufficiency of mortar for all those works, the whole or a part of which is under the water, induced the scientific builder and chemists to seek a substitute. Many compositions have been recommended, and several of them have been found to answer the purpose. There is one substance however, Roman cement, which, above all others, is extremely useful for a number of purposes, and will require our attention; and if our remarks should occupy a space which may appear to have no proportion to the length of the other parts of the volume, the importance of the subject will be a sufficient excuse.

Roman Cement.

Roman Cement was accidentally discovered in the year 1796, by Mr. Parker, whose attention was attracted, when walking beneath the cliffs of blue clay, on the shores of the island of Sheppy, by the uniform appearance of the masses of stone which were strewed here and there upon the beach, and were seen projecting from the cliffs. As a mere matter of curiosity, he collected two or three fragments, and happened afterwards to throw one of those pieces into the fire. After it had been exposed for some time to the fire, it fell upon the hearth, and was there found by Mr. Parker, who was induced to make some experiments upon its cohesive properties, which led him to the discovery of its value, as a strong and durable cement. He then immediately applied to the government of the day, for a patent, which was granted to him, for fourteen years; and having secured to himself the exclusive right of manufacture, realized an ample fortune.

So great has been the recent demand for cement stone, that its quantity has been much diminished, and other substances have been substituted to so great an extent, that the cement now used is much inferior to that originally manufactured by Mr. Parker. So small is the quantity obtained on the Sheppy coast, that the manufacturer is scarcely repaid for the cost of a search. The natural physical causes which are constantly active, have a tendency to increase the quantity upon the beach which surrounds this interesting island; but all natural agents act in a slow and progressive manner, so as to afford a very inadequate supply for the demand which is now made for this material. The masses once abundantly strewed over the shores of the island of Sheppy have been long since removed by the cement manufacturers, and the supply which is now obtained from this spot depends upon the quantity of the cliff that may be thrown down, by the undermining influence of land springs, or by other causes. At the base of the cliffs which surround this island may be seen, here and there, extensive land springs, which weaken the foundation of the clay, and frequently cause masses of large extent to fall upon the beach. This cause is aided by the storms which, during the winter season, frequently blow upon its shore, and either by the force of the waves or by the subsequent drying of the saturated mass of clay, weaken its cohesion, and produce the same effect. The falling of the cliffs, produced by these and other means, furnishes a small quantity of cement stone, but a quantity altogether inadequate for the supply of the demand. But as far as observation extends, it appears that these nodular stones are found in all the deposits of London or blue clay. This stratum is found in Harwich, and other places, as well as at Sheppy; and the attention of the manufacturer was consequently directed to them, for a supply of the material. But it has been stated, and experiment seems to justify the assertion, that the Sheppy stone yields a much better cement than that which is obtained from other places; the cause of this cannot be readily determined; but so great a value is placed upon the former, that some persons have actually excavated for the purpose of obtaining it. But the principal part of that now used by manufacturers, is obtained from Harwich; and not less than from thirty to forty tons weight are annually collected in this place. The engineer and architect still prefer the Sheppy cement, which has a much lighter colour than that made from the Harwich stone, but is far more expensive. The manufacturer, however, now so frequently intermixes other ingredients with the Harwich cement, to give it the same appearance as the Sheppy, that it is almost impossible to determine the quality by the colour. Limestones, found in other places, have been substituted for the Sheppy nodules; all of which, excepting that which is found in small quantities in the marshes of Essex, near Steeple, are much inferior to it.

The manufacture of cement is extremely simple, although some experience is necessary, as the character of the cement will depend as much upon the manner in which it is made, as upon the property of the stone. After the stone has been broken into small pieces it is thrown into a kiln, with a proportion of small coal, to be burnt. A strong red heat must now be supported throughout, and considerable skill, or rather experience, is required to accomplish this purpose; for the relative degrees of hardness in the several pieces, and other causes, tend to give them an unequal temperature, and to prevent perfect calcination. After the stone has remained from thirty to forty hours in the kiln, in which time it is usually perfectly burnt, it is taken to the mill, and being immediately ground to powder, is packed in casks and sent into the market. Promptness in all the processes which follow burning is absolutely necessary, for the contact of the air impairs the adhesive power of the cement. Hence it is that builders, who study the character of their materials, invariably prefer the cement which is made in large manufactories; a ready sale generally securing them from the use of an old cement. Good cement perfectly burnt has a light-brown colour, and has very little weight; but if imperfectly burnt it is heavy and has a dark colour. When the stone is burned overmuch, small black carbonized particles may be observed. It may be necessary to state that the cement should always be reduced to as fine a powder as possible; and to accomplish this an attempt was made some time since to sift it, but its exposure to the air was found to injure its properties as a cement. As a test of the value of a cement, the experimenter may mix with it a quantity equal to two-thirds of clean, well washed, and dried sand, and should it then have a strong cohesive power, he may depend upon its qualities; but as soon as the two ingredients are mixed and moistened, the cement should be used, or it will either fail to set, or to possess an adequate cohesive power. These suggestions, if carried out, will be found of great importance in the art of building, and particularly in those instances where great stability is required. The builder frequently attributes to the cement that which depends upon his own injudicious use or exposure of the material; and even bad cement may be made tolerably effective for ordinary purposes, if it be little exposed to the atmosphere, and be used immediately after its mixture.

Roman cement should never be used in any situation where there is the slightest chance of warping or spring, for as it does not possess any elastic force it is sure to break away. For covering walls, when used as a stucco, it is well suited, but the bricks should be damped previous to its application, or they will absorb its moisture and give it a porous structure. But stucco will not bind upon a bed of stucco, and it is therefore necessary that it should be applied in one coat: for, as good cement will set in about twenty minutes, a second bed cannot be applied at any subsequent period without endangering the stability of the work, for one coat is almost sure to separate from the other.

D

To ascertain the relative value of any number of cements, mix them with certain porportions of sand, and that which is the hardest with the largest proportion is the best. As a collateral proof, the specimens may be kept for a few days, and it will be found that the quantity of bloom formed upon their surfaces will have some relation to their qualities. Good cement will always be raised to a great temperature when mixed, and those which are not may be considered worthless. There are some cements that harden very quickly, and yet are of very bad quality, and will in the course of a few hours become quite soft. These facts are well worthy the attention of the workman or the builder, for they will not only enable him to ascertain which is a good and which a bad material, but also to use the material he may choose, in the most advantageous manner.

Chemists and others, who have investigated the properties of hydraulic limes, are not by any means agreed as to the cause of the cementitious quality. Saussure was of opinion that their peculiar properties were derived from the presence of silex and alumina in certain proportions; Descotils attributes them to the presence of a large proportion of silex, and Bergman and Guyton to a small proportion of manganese.

The Roman is the most valuable of all water cements, as well for the ease with which it may be used, as for its hardness and durability. As it sets in about fifteen minutes, the workman cannot mix more than a small quantity at once. Experience will soon teach how much can be worked in a certain time: an appropriate quantity must be taken upon a clean board, and something more than an equal quantity of very clean and dry river sand. When the lime and sand are thoroughly mixed, as much clean water as is necessary to form them into a paste should be added. The workman should then immediately use it, and after it has been once applied, it should not be in any way disturbed. Forty bushels of cement, with its appropriate quantity of sand, will do a rod of brick-work. Good cement will take two parts of sand, and that cannot be called good which will not take one and a half.

When cement is used for coating or lining walls, it must have as much sand as possible, so as not to be too stiff to work. It must also be always worked in one coat, and the surfaces to which it is applied should be clean and well wetted. Cement when thus used is called stucco, and should be laid on three-quarters of an inch in thickness. A bushel of cement with its proper proportion of sand will cover a surface of two square yards.

Cement is also used for casting ornaments, for which purpose it answers exceedingly well. Gothic work is sometimes finished in this way, but, although it may be desirable in some instances, it is generally better to use stone, where very ornamented work is to be introduced.

There are several other kinds of cement which are occasionally employed by the bricklayer, but they are not of sufficient importance to be treated of, in a work which can only give some of the most prominent facts in the art of building. But it may be asked, what was used before Parker's cement was discovered? This question leads us to make a few remarks upon two cements which were once extensively used in this country, Puzzolana and Tarras, but which are now scarcely ever employed.

Puzzolana.

Puzzolana is a volcanic substance, consisting, according to Bergman's analysis, of from fifty-five to sixty per cent. of silica, from nineteen to twenty of alumina, five of lime, and twenty of iron. The Romans were accustomed to mix this substance with lime, in the manufacture of water-cements, and the same method was a long time adopted in England. The hardening of the mortar is supposed to arise from the union of the oxygen of the water with the iron.

Tarras.

Tarras or Tras is a substance found at Andernach, in the department of the Rhine, and, according to Bergman, differs but little from Puzzolana in composition. Tarras mortar is well suited for all those situations in which it is constantly exposed to water, but it cannot resist the action of alternate wet and dry, and indeed is never so firm when it sets in exposure to air, as under the water. The principal objection to the use of this mortar was its expense, and consequently the Dutch attempted to supply its place by the union of substances found in their own country; they succeeded so well, that a large quantity was imported into this country,* and extensively used. There are two proportions which have been adopted as the best for the tarras mortar; in one kind a measure of quick lime is mixed with a measure of Tarras, and, being thoroughly mixed, are brought into the consistence of paste by the addition of water, as little water being used as possible; in the other, one measure of Tarras is mixed with two measures of slaked

* England.

lime and three of sand,—this is almost as good a cement as the former, and much cheaper.

THE METHODS OF LAYING BRICKS.

The strength of walls and piers of brickwork depends as much on the manner in which the courses are placed, as on the quality of the materials employed in construction; for, however good the bricks may be, if they are not so placed as to strengthen one another, and mutually confine each other to their several situations, the work cannot have the requisite stability. If the perpendicular joints in the several courses are too nearly over each other, the work is liable to crack in a vertical direction, and if the bricks, forming the outer and inward face of the wall do not bind together, the work will bulge, and the wall must at last fall to pieces by its own weight. It is therefore important for us to determine the best method of laying bricks, and we shall endeavour to describe the means adopted by builders to prevent the separation of the work, and give a solid bearing to every part.

Those bricks which are so placed that their length is in the direction of the wall, are called stretchers; and those which are placed with their length across the wall, are called headers.

The two principal methods of bricklaying are severally called English and Flemish bond. English bond is generally preferred by builders, as being decidedly the strongest, though it has not so neat and regular an appearance as Flemish. English bond consists of alternate courses of headers and stretchers; thus, one course is formed with headers, that is, with bricks crossing the wall; the next with stretchers, that is, with bricks having their length in the same direction as that of the

D *

wall: the headers serve to bind the wall together in a longitudinal direction, and the stretchers prevent the wall from separating crossways.

Flemish bond consists in placing a header and a stretcher alternately throughout every course. This method of bricklaying is very much adopted, on account of the regular appearance it gives to the face of the work, but, in order to have this result, a header must always be placed over the middle of the stretcher below it. The Flemish bond, though inferior in many respects to the English, is very generally used, and an inferior brick is placed in the interior of the wall, and those which form the face, are picked or chosen, that the work may have a uniform colour. The greatest fault in this method of bricklaying is, that by making a putty joint on the face, the interior bricks do not range level with the exterior ones, and this prevents the builder from connecting his work by headers extending through the whole thickness of the wall.

THE CARPENTER.

A CARPENTER is a workman who executes that combination of timbers which may be considered, in connexion with the bricklayer's work, as the frame or skeleton of a building. There is, however, this difference between the objects of the one and of the other; the bricklayer has only to consider the downward pressure, or force of gravity, and the forces which may be exerted to destroy the perpendicular; the carpenter must also study the relative disposition of parts, so as to alleviate as much as possible the strains which may be exerted on the building.

Carpenter's work is distinguished from that of the joiner; for while the one has regard to the substantial parts of an edifice, or those which give solidity and strength, such as the construction of roofs, floors, and partitions, the other consists in providing for the ornamental and convenient. A carpenter should be well acquainted with the strength and character of the materials he uses, and especially as he employs them in great masses. He should also be careful not to overload a building, or to employ larger timbers than are absolutely necessary; for, even if there were no danger in so doing, economy would dictate the necessity of this care. It is then important that the carpenter should be able to ascertain the dimensions required for the several parts of a building, so as to produce a maximum of strength, without overloading the walls or his own work, and at the same time, to avoid the danger which must result from a scantiness of material. There are then two things to be considered, the strength of the materials, and the stress to which they are subject in certain situations. A timber, or framing, may be strained in various ways, but of these we shall speak presently; our first object is to describe the materials themselves, referring particularly to those woods which are most commonly used.

Oak.

There are many species of oak, but that known among botanists, as the "Quercus robur," is most esteemed. It may, however, be necessary to remark in relation to this, as well as all other kinds of timber trees, that the character of the wood must greatly depend upon the soil in which it grew, and the degree of attention it received from the cultivator. The oak of Sussex is most esteemed by builders, but, whether the preference is dictated by experience or prejudice, we are unable to state; but we are not acquainted with any series of experiments that warrant the choice, and it is not fit that practice should be regulated by unproved statements.

A Norway oak, called clapboard, is frequently brought to London; and also one that is grown in Germany, called Dutch wainscot, being imported from Holland, to which country it is brought in floats down the Rhine. Both these woods have been extensively used in this country, and it is probable that the wainscot will be still employed for many purposes, for, though it is softer and the grain more open than the English oak, it is also less liable to warp.

Oak is the most durable of all woods, and surpasses them in strength and stability. Vitruvius says, that it has an eternal duration; and when we see the beautiful specimens which have remained untouched by time, in our oldest buildings, though all other materials are crumbling around them, we feel an inclination to assent to his opinion. They are, however, only the close grained varieties that deserve this character; and it is no small addition to the professional skill of the architects of past ages, that, by the choice of the best materials, they gave a perpetuity to their works, which few, if any, of the present day can rationally expect.

Oak may be used in all those places where strength is required, and its flexibility does not present an objection. For sleepers, wall-plates, ties, king-posts, and other such purposes, it should be used more frequently than it is. But its chief application is for ship-timber, and some thousand loads are annually used in our Dockyards. This remark suggests the propriety of using it in all those places which are much exposed to the variation of weather.

Fir.

There are many species of fir, all of which are more or less used in building; but there are three sorts in particular that require our attention, being more used than any others : these are the Pinus Sylvestris, or yellow fir; the Pinus Abies, or spruce fir; and the Pinus Resinosus, or pitch pine.

The red or yellow fir is a native of Scotland, and the Northern countries of Europe. This tree is more abundant than all others in the boundless forests of Norway and Sweden. It grows to an immense height, very straight, and with few branches. The fir timber of Norway is brought into this country under the name of masts and spars ; those which are eighteen inches or more in diameter are called masts, and are frequently eighty feet in length ; others are called spars. In several parts of Scotland the yellow fir is grown, and attains a great height.

The yellow fir or deal is much used in building, and is

a very durable wood; according to some authors, as much so as oak. But whether this be the case or not, it has many qualities which render it exceedingly useful to both the carpenter and the joiner. It is light, and easily worked, yet stiff, and capable of bearing great weights. It is commonly employed for framing, girders, joists, and·rafters; for joiner's work also it is almost universally used.

White fir is also a native of the north of Europe; it is especially abundant in Norway and Denmark, and is sometimes called the Norway spruce. The larger quantity of that which is brought into this country, is imported from Christiana in deals and planks. Deals are formed by cutting the fir tree into thicknesses of generally about three inches, the width being about nine. As fir is exceedingly liable to shrink, it is very necessary that it should be well seasoned, and this is especially the case with white fir, which should never be used in those places that are exposed to atmospheric changes. We are informed by travellers, that the tree is first cut into three lengths of about twelve feet long, each of which is divided into three deals.

The pitch pine, which is a native of Canada, is sometimes employed by the carpenter, but not so frequently as those kinds we have already mentioned. This wood is much heavier than either of those we have already described, but it is less durable. Its name has been derived from the circumstance of its containing a large quantity of resin, which makes it very unfit for building purposes, and very difficult to work.

Larch.

· There are three species of Larch; one is a native of Germany and the neighbouring countries, the other two are Americans. The European species (Pinus larix,) sometimes grows to a great height, and contains a large quantity of timber; one, which was cut at Blair Atholl in 1817, is said to have contained 252 cubic feet of timber; this however, was a tree of remarkable size.

Mr. Tredgold, in his most interesting and useful work on Carpentry, has made some appropriate remarks upon the character of this wood. "It is extremely durable in all situations, failing only where any other kind would fail: for this valuable property it has been celebrated from the time of Vitruvius, who regrets that it could not be easily transported to Rome, where such a wood would have been so valuable. It appears, however, that this was sometimes done; for we are told that Tiberius caused the Naumachiarian bridge, constructed by Augustus, and afterwards burnt, to be rebuilt of larch planks, procured. from Rhœtia. Among these was a trunk 120 feet in length, which excited the admiration of all Rome. The celebrated Scamozzi also extols the larch for every purpose of building, and it has not been found less valuable, when grown in proper soils and situations, in Britain. In posts and other situations, where it is exposed to damp and the weather, it is found to be very durable. In countries where larch abounds, it is often used to cover buildings, which, when first done, are the natural colour of the wood, but in two or three years they become covered with resin, and as black as charcoal; the resin forms a kind of impenetrable varnish which effectually resists the weather. Larch is not attacked by common worms, and does not inflame readily.

The larch is useful for every purpose of building, whether external or internal; it makes excellent ship-timber, masts, boats, posts, rails, and furniture. It is peculiarly adapted for flooring-boards, in situations

where there is much wear, and for staircases; in the
latter, its fine colour, when rubbed with oil, is much
preferable to that of the black oaken staircases to be seen
in some old mansions. That we may not give an erroneous estimate of the value of the larch, as applicable
to building purposes, it is necessary to state that it is
worked with more difficulty than fir, and is even more
liable to warp, unless it be perfectly seasoned.

Beech.

The beech (Fagus sylvatica,) is not much used in
building, on account of the very rapid decay which it
undergoes, whenever it is affected by dampness. It
grows in our own, as well as in most European countries; but it prefers a dry soil, and, in England, flourishes
most in chalk districts. There are two kinds of beech-
wood; one is called the brown or black beech, the other
the white; it is, however, generally supposed that the
difference is due to the character of the soil, and not to
any specific distinction. Beech is a hard, fine-grained
wood, and has been much used for the commoner kinds
of household furniture. It may appear singular that it
should be well adapted for piles, provided it is constantly immersed in water; but damp destroys it very
readily. Nor is this the only objection to its being used
in building; for even the best, which is the white, is
soon injured by worms, whether in a dry or a damp
situation.

Ash.

There are several species of ash, but the one which is
most common in Europe, called by botanists the Fraxi-
nus excelsior, is the most valuable. The tree sometimes

grows to an immense size; but its mean diameter is said not to exceed twenty-three inches. The texture of the wood is alternately compact and porous, and presents a veined appearance, the veins being darker than those of the oak. On account of its great flexibility, and want of durability, it is never applied for framing or for timbers. From the experiments which have been made, it appears that it is tougher and stronger than oak, and, were it not for its great flexibility, might be, in many instances, advantageously employed by the carpenter. It is not, however, without a use in the arts, being exceedingly well adapted for many parts of machines and carriages.

Elm.

Five species of elm are found in this country; but the wych elm (Ulmus campestris,) and the smooth-barked elm (Ulmus glabra,) are most valuable. Elm decays rapidly when exposed to variations of weather; but is durable when kept constantly dry, or constantly under water. The piles upon which Old London Bridge was erected, were elm, and their soundness, after an exposure to water for some centuries, proves the truth of one of these statements. It is a porous and generally coarse cross-grained wood; and, on this account, should never be used in any piece of framing where a strain is to be supported. But, in addition to this, it is liable to shrink both in breadth and length, though it is not readily split. It is by no means an important wood to the builder; but a large quantity is used in this country. For many hydraulic works it is very useful; some parts of ships are constructed of it; and it is generally employed for coffins, piles, and wet planks. The wood of the wych elm is preferred to all others.

E

Chesnut.

The chesnut (Fagus castanea,) is one of the most long-lived of all European trees. It is a native of many parts of Europe, and was at one time very common in England, yielding the principal timber at the time. The roof of King's College, Cambridge, is made of chesnut, which is one instance of its durability in a dry state. It is also well adapted for water-pipes, casks, and other vessels intended to hold fluids. When thoroughly seasoned it will neither shrink nor swell, and may be applied for all those purposes for which oak is used, and in some instances is more useful. The wood is hard, and, when young, tough and flexible. It is not always easy to distinguish between oak and chesnut, for they much resemble each other in colour and in grain; but they may be known, says Sir Humphrey Davy, " by this circumstance, that the pores in the alburnum of the oak are much larger and more thickly set, and are easily distinguished; while the pores in the chesnut require glasses, to be seen distinctly." The wood of old trees is generally brittle, and should never be used in those situations where it will be subject to a considerable strain. It has also been stated, that when chesnut is shut out from the access of air, it quickly decays. It is much to be regretted that the culture of this tree, at once ornamental and useful, should be so much neglected in England. In some instances it has been known to live from eight hundred to a thousand years; and its full and beautiful foliage might induce the land proprietor to propagate it, even if he should be uninfluenced by its usefulness in the art of building.

Walnut.

The common walnut (Inglana regia,) is a native of Persia; but was once cultivated in this country as much for its wood as its fruit. It is a greyish-brown wood, with a fine grain; but, if it were not scarce, and could be obtained by the builder for the same money as the woods now employed by him, it would be very unfit, on account of its flexibility and aptness to split, for any of those situations where a weight is to be sustained; though it was sometimes used for this purpose in former times. It is now chiefly used for gun-stocks, handles to steel instruments, and for furniture. It is less liable to be attacked by worms than perhaps any other wood, excepting cedar. For some building purposes, particularly for some joiner's work, it might be advantageously employed, could the supply be sufficient.

Mahogany.

This wood is the produce of a tree called the Swietenia mahogani. It is much used by cabinet makers, and frequently by joiners for doors, hand-rails, tops of counters, and other ornamental work. The tree is a native of the West India Isles, and of the Bay of Honduras. On account of its costliness, it cannot be extensively used in this country by the carpenter, though its qualities are such as would make it otherwise desirable. The Spanish mahogany, or that which grows in the West Indies, is most esteemed, and is imported in lengths of about ten feet, and from twenty to twenty-six feet square.

Teak Wood.

Teak wood, or Indian oak, is obtained from the Coromandel coast. It is a light and durable wood, easily worked, and equal if not superior to oak in strength and stiffness. It is chiefly used for ship-building; a purpose for which it is well adapted, being of an oily nature, and yielding good tar.

Poplar.

Several kinds of Poplar grow in England, but none of them are much employed by builders. The wood has a beautifully clean grain; it is light, though not very strong; is easily worked; and may be sometimes used for flooring in those situations where there cannot be much wear.*

* To the above may be subjoined the following, which are not found in the English edition of this work, as not being much in use there. The accounts are chiefly from the "Library of Entertaining Knowledge."

Maple.—Of this useful timber there are about twelve species indigenous to America, the principal of which, for the purposes of the carpenter and joiner is, the *Great Maple* (*Acer pseudo-platanus*). The great maple, called also the sycamore and the plane tree, is hardy ; stands the salt spray of the sea better than most trees ; grows rapidly, and to a great height. The timber is very close and compact, easily cut, and not liable either to splinter or warp. Sometimes it is of uniform colour, and sometimes it is very beautifully curled and mottled. In the latter state, as it takes a fine polish, and bears varnishing well, it is much used for certain parts of musical instruments. Maple contains none of those hard particles which are injurious to tools, and is therefore employed for cutting-boards ; and not being apt to warp, either with variations of heat or moisture, it is an eligible material for saddle-trees, wooden dishes, founders' patterns, and many other articles both of furniture and machinery. Before the general introduction of pottery-ware, it was the common material for bowls and platters of all sorts ; and many are still made of it. If the timber be placed so that insects are allowed to settle upon it, it is speedily attacked by the worm. When kept dry and

The woods we have described are the most important of those used by the carpenter and joiner. To distinguish the one from the other, the reader must accustom himself to examine specimens carefully ; for it is impossible, by any description, to give him a capability of doing so. Our object has been to relate the characters and properties of the several kinds of timber, as deduced from the experiments which have been made by practical and scientific men. There is one thing, we think, that will particularly strike the reader's attention, and should be constantly borne in mind : the same wood is not equally useful in different circumstances ; and when we discover that it possesses durability in one situation, it by no means follows that it will have the same property in another. A wood may be admirably suited for floors, but it may be altogether unsuited for timbers, or for situations where great weights are to be sustained.

DECAY OF WOOD.

Allusion has been frequently made in the preceding remarks to the fact, that wood is under some circumstances, susceptible of decay. Some woods decay much

---

free from this attack, it will last a considerable time ; but, exposed to humidity, it is one of the most perishable of trees.

The White Walnut, or Hickery.—This wood is a native of North America, where it grows to a considerable size. One part of it is more porous than that of the walnut, but the other is more compact ; this gives the grain of the wood somewhat the appearance of that of ash. Hickery is very tough and elastic ; and therefore it answers remarkably well for fishing-rods, the shafts and poles of carriages, and other purposes where a slender substance of timber has to resist sudden jerks or strains.

Fancy Woods.—As these more particularly appertain to the business of the cabinet-maker, it may be sufficient, for the purpose of making the catalogue of woods more complete, to name them. The principal are the following, which are commonly cut into veneers from nine to twelve or fourteen to the inch. *Rose-wood, King-wood, Bas-wood, Tulip-wood, Zebra-wood, Satin-wood, Sandal-wood, Amboyna-wood, Ebony, Snake-wood, &c.*

E*

more rapidly than others; but they will all, in some situations, lose their fibrous texture, and, with it, their properties. But all circumstances are not equally conducive to decay; for it will be evident that there must be some arrangement of causes to produce this effect. To ascertain the causes which act upon woods, and effect their destruction, is an important object both to the builder and to the public; for, until this has been done, we cannot ever expect to ascertain any general principle, that may guide us in our endeavour to avoid those circumstances which have a tendency to encourage the destruction, or to propose a remedy for the evil. The ravages which are constantly made upon all our works of art, give a character of insecurity to our labours; for the things which men accomplish with great perseverance and difficulty, in a length of time, are in a few years, destroyed by invisible agents. In studying the decay of wood, there are three things that demand our attention, the causes, the circumstances under which those causes are most active, and the means by which they may be destroyed, or their effects in some degree neutralized.

CAUSE OF THE DECAY OF TIMBER.

All vegetable as well as animal substances, when deprived of life, are subject to decay. From a very early period attempts have been made to prevent this decomposition; and in some degree these attempts have been successful, more especially with animal bodies. The Egyptians were acquainted with so perfect a means of embalming animal substances, that the bodies of men and animals, prepared by its earliest inhabitants, have combated for centuries the influence of time, and have been found in a perfect state by our contemporaries. This being effected, it is reasonable to hope that some means may yet be provided, that shall arrest the destruction of vegetable substances. It is not to be expected that it will ever be possible to give a perpetuity to a particular form of substance, but it is possible to remove in part the cause, and thus to give a lengthened continuance to one particular constitution of elementary principles.

If the trunk or branch of a tree be cut horizontally, it will be seen that it consists of a series of concentric layers, differing from each other in colour and tenacity. In distinct genera, or species of trees, these layers present very different appearances, but in all cases the outer rings are softer and more porous than the interior. Wood is essentially made up of vessels and cells, and the only solid parts are those coats which form them. These vessels carry the sap, which circulates through the tree, gives life and energy to its existence, and is the cause of the formation of leaves, flowers, and fruit. But when the tree is dead, and the sap is still in the wood, it becomes the cause of vegetable decomposition by the process of fermentation. Fourcroy, the celebrated chemist, says, there are five distinct species of vegetable fermentation, the saccharine, the colouring, the vinous, the acetous, and the putrefactive. We are but little acquainted with the process by which the decomposition is carried on; but the effect is certain, unless the albumen, one of the constituent proximate principles of vegetable matter, be disposed of, or be made to form with some other substance a compound not subject to the same process of decay. We are, it appears, indebted to Mr. Kyan for the discovery that albumen is the cause of putrefactive fermentation, and the subsequent decomposition of vegetable matter.

Circumstances favorable to Vegetable Decomposition.

Wood is not equally liable to decay under all circumstances. When thoroughly dried it is not so quickly decomposed as when in its green state, for in the latter condition it has in itself all the elements of destruction, and it is scarcely possible to prevent the effect, if it be then used in building. But supposing the timber to be perfectly seasoned, it is more liable to decay under some circumstances than in others. Timber is most durable when used in very dry places. Time, however, which decays all material things, affects the hardest wood, even when employed in the most advantageous circumstance. Yet timber, which has been used in places where it receives no other moisture than that which it absorbs from the atmosphere, has been known to last for seven or eight hundred years, though its elastic and cohesive powers are invariably injured.

When timber is constantly exposed to the action of water, the decomposition effected will depend upon the nature and chemical composition of the substance. Vegetable matter is a compound, and an ingredient may be removed without destroying the whole. A portion of the wood may be soluble in water, but other parts are not; so that after a definite period, the continued action of water upon a piece of timber ceases, and if it can sustain the influence of this cause until that period, there is no termination to its endurance, except from those casualties which it might have been able to bear in its original state, but cannot after the removal of that portion of its substance soluble in water. Should a piece of timber, that has been for a long time exposed to water, be brought into the air and dried, it will become brittle and useless; this is usually the case with the timber taken from peat bogs, unless it should happen to be impregnated with some mineral substance that has stayed the action of the water.

When wood is alternately exposed to the influence of dryness and moisture, it decays rapidly. It appears, from experiments that have been made, that after all the matter usually soluble in water has been removed, a fresh maceration and contact of the air produces a state of matter in that which is left, which renders it capable of solution. A piece of timber may then in this manner be more and more decomposed, until at last the whole mass is destoyed. The builder is sometimes compelled to use wood in places where it will be exposed to alternate dryness and moisture; fencing, weatherboarding, and other works, are thus exposed. In all these cases he may anticipate the destructive process and provide against it. The wood used in such situations should be thoroughly seasoned, and then painted or tarred; but, if it be painted when not thoroughly seasoned, the destruction will be hastened, for the evaporation of the contained vegetable juices is prevented.

There is one other circumstance to be considered, the influence of moisture associated with heat. Within certain limits the decomposition resulting from moisture increases with the temperature. The access of the air is not absolutely necessary to the carrying on of this process, but water is; and as it goes on, carbonic acid gas and hydrogen gas are given off. The woody fibre itself is not free from this decomposition, for as the carbonaceous matter is abstracted by fermentation, it becomes more susceptible of this change. This statement is proved by the circumstance that, when quick-lime is added to the moisture, the decomposition is accelerated, for it abstracts carbon. But the carbonate of lime produces no such effect: a practical lesson may be learnt from this fact; if timbers be bedded in mortar, decomposition must follow, for it is a long time before it can absorb sufficient carbonic acid to neutralize the effect; and the dampness which is collected by contact with the wet mortar increases the effect. When the wood and the lime are both in a dry state, no injury results, and it is well known that lime protects wood from worms.

When the destructive process first becomes visible, it is by the swelling of the timber, and the formation of a mould or fungus upon its surface. This fungus or cryptogamic plant rapidly increases, and soon covers over the whole surface of a piece of timber, having a white, greyish-white, or browish hue. When the seeds of destruction are thus once sown, they cannot be readily eradicated; it need not therefore be a matter of surprise that many of the foreign woods used in this country have so little perpetuity, when the reader is informed, that the heat of the hold of the vessel in which they are brought is sufficient of itself to cover them with mould or mildew. Heat and moisture may be considered the prominent causes of the rapid decomposition of vegetable substances. When wood is completely and constantly covered with water this effect is not produced, and we have an example in the fact, that, although those parts of a vessel which are subject to an occasional moisture are liable to dry rot, yet those parts which are constantly beneath the water are never thus affected; and although the head of a pile, which may be now and then wetted by the casual rise of the time, and is then dried again by the sun, may be decomposed, yet those parts which are always covered with water have been found in a solid state after centuries of immersion.

MEANS OF PREVENTING DECAY.

It cannot be thought a matter of small importance to have some means of preventing the decay to which wood appears to be so subject. Many experiments have been made, under the hope of discovering a simple and effective process for the accomplishment of this purpose. Whenever there is a desirable object, which seems to offer a prospect of fame or wealth to him who can secure it, there will always be many persons who, impelled by a sanguine disposition, or by bad motives, will propose schemes which are not founded on scientific principles, and frequently produce more harm than good. This we have frequently seen, and in a time like the present, when too many seem to be speculating for an existence, rather than seeking wealth and honourable independence by the legitimate exertion of intellect or skill, the public are peculiarly exposed to the impositions of the weak and of the crafty. Scarcely a month elapses but we hear of some new specifics against the decay of timber, and yet, when brought to the test of experiment, they are found to be utterly useless. Some fortunate observations, some unexpected result, as the patentees inform us, led to the discovery; and as to the reason why this or that process should be effective, they neither know nor care. We do not, however, in these censures include the process proposed by Mr. Kyan, which we shall presently have occasion to explain.

Felling Timber.

Something may be done towards the prevention of decay, by felling the timber at a proper season. A tree may be felled too soon or too late, in relation to its age, and to the period of the year. A tree may be so young that no part of it shall have the proper degree of hardness, and even its heart-wood may be no better than sap-wood; or a tree may be felled when it is so old that the wood, if not decayed, may have become brittle, losing all the elasticity of maturity. The timber grower is more likely to adopt, from interested motives, the former of these errors, and fell his timber too young. His object is to obtain as much timber as possble, but a tree is not in its maturity when it ceases to grow, for after this period its fibres gain firmness and density. The time required to bring the several kinds of trees to maturity varies, according to the nature of the tree and the situation in which it may be growing. Authors differ a century as to the age at which oak should be felled, some say one hundred, and others two hundred years; it must therefore be regulated according to circumstances. Although the oak of our own country is so valuable to the builder, yet it is to be feared that it is seldom allowed to attain its maturity, the grower being anxious to sell, and the builder to buy; the one seeking to obtain its value himself, rather than leave it to posterity, the other to purchase at as low a price as possible, not caring sufficiently for the character of the timber.

But it is also necessary that the timber-trees should be felled at a proper season of the year; that is to say, when their vessels are least loaded with those juices which are ready for the production of sap-wood and foliage. The timber·of a tree, felled in spring or in autumn, would be especially liable to decay; for it would contain the element of decomposition. Mid-summer and mid-winter are the proper times for cutting, as the vegetative powers are then expended.

There are some trees, the bark of which is valuable, as well as the timber; and as the best time for felling is not the best for stripping the bark, it is customary to perform these labours at different periods. The oak-bark, for instance, is generally taken off in early spring, and the timber is felled as soon as the foliage is dead; and this method is found to be highly advantageous to the durability of the timber. The sap-wood is hardened, and all the available vegetable juices are expended in the production of foliage. Could this plan be adopted with other trees, it would be desirable; but the barks are not sufficiently valuable to pay the expense of stripping.

Seasoning Timber.

Supposing all these precautions to be taken in felling timber, it is still necessary to season it; that is, to adopt some means by which it may be dried, so as to throw off all the juices which are still associated with the fibres of the wood. As soon as the timber is felled, it should be removed to some dry place; and, being piled in such a manner as to admit a circulation of air, remain in log for some time, as it has a tendency to prevent warping: The next process is, to cut the timber into scantlings, and to place these upright in some dry situation, where there is a good current of air, avoiding the direct rays of the sun. The more gradually the process of seasoning is carried on, the better will be the wood for all the purposes of building. Mr. Tredgold says, "It

F

is well known to chemists, that slow drying will render many bodies less easy to dissolve; while rapid drying, on the contrary, renders the same bodies more soluble. Besides, all wood, in drying, loses a portion of its carbon, and the more in proportion as the temperature is higher. There is, in wood that has been properly seasoned, a toughness and elasticity which is not to be found in rapidly-dried wood. This is an evident proof, that firm cohesion does not take place when the moisture is dissipated in a high heat. Also, seasoning by heat alone, produces a hard crust on the surface, which will scarcely permit the moisture to evaporate from the internal part, and is very injurious to the wood.

"For the general purposes of carpentry, timber should not be used in less than two years after it is felled; and this is the least time that ought to be allowed for seasoning. For joiners' work it requires four years, unless other methods be used; but, for carpentry, natural seasoning should have the preference, unless the pressure of the air be removed."

Many artificial methods of seasoning timber have been proposed; a brief notice of some of those which have been found most useful will be required.

Seasoning by a Vacuum.

All the vegetable and animal juices are kept in their particular vessels by the pressure of the atmosphere; remove that pressure, and the animal fluids could no longer be retained by the veins and arteries, and the vegetable fluids would exude and appear on the surface of the plant. Place a small piece of wood beneath the receiver of an air-pump, and exhaust the air, and in a short time the wood will be covered with drops of the liquid which can no longer be retained, as the atmospheric pressure is removed. Mr. Langton thought that this might be applied to the extraction of those vegetable juices in timber, known to be the cause of its decay. An arrangement was therefore adopted, by which large masses of timber might be enclosed in a vessel having such machinery as would be necessary to exhaust the air, heat being at the same time employed so as to vaporize the exuded juices. The vapour is conveyed away by pipes surrounded by cold water, and is condensed into a liquid, having a sweet taste. This process is deserving of more attention than has hitherto been given to it.

Water Seasoning.

It has been stated, by various writers, that wood immersed in water for about a fortnight and then dried, is better suited for all the purposes of the joiner. There can be no doubt that immersion in water tends to neutralize the effect of the saccharine matter, by dilution or an almost absolute removal. This process has also the effect of rendering the wood less liable to crack and warp; but, if we judge by Duhamel's experiments, it injures the strength of the material, and should not, therefore, be adopted in any instance where the timber is to be employed by the carpenter. Evelyn recommends boards that are to be used for flooring, to be seasoned in this way: "Lay your boards," he says, "a fortnight in water (if running, the better, as at a millpond head;) and then setting them upright in the sun and wind, so as it may pass freely through them, turn them daily; and thus treated, even newly-sawn boards will floor far better than those of a many years' dry seasoning, as they call it." Timber intended for ship-building may be immersed in sea-water; but that which is to be used for houses ought to be placed in fresh water; for if timber, or any other building material, be impregnated with salt, it will ever be wet; for, salt attracts moisture so readily, that it may be used approximately as a hygrometer. Plaster or mortar made with salt water, will always sweat with a moist atmosphere; and timber intended for the house-carpenter, if impregnated with salt, will always be damp, or covered with a crystalized efflorescence. Much injury, however, is sometimes done by not thoroughly immersing the timber; the carpenter should therefore be careful, when he employs this method of seasoning, that the timber be entirely covered with water, and that it be not exposed to its action for too long a time.

Seasoning by Smoking and Charring.

Authors, who have written upon the seasoning of timber, have spoken of the effects of smoke, and the carbonization of the surface. We have adopted the same arrangement, but it will be necessary to caution the reader against a misconception of a very inaccurate expression. Timber cannot be seasoned either by smoking or charring, but seasoned timbers may be made more capable of resisting the effects of certain situations, by these processes. Should a piece of timber, containing the vegetable juices, be smoked or charred, it would be a means of accelerating decomposition; for, preventing all means of evaporation, the common sources of protection would become sources of destruction. But when timber is to be used in situations where it is liable to be attacked by worms, or to produce fungi, it may be desirable to smoke or to char it.

Seasoning by Boiling or Steaming.

Timber is sometimes seasoned by steaming or boiling, both of which means are frequently adopted by shipbuilders. The strength of timber appears to be somewhat impaired by these processes, but it is generally less liable to shrink or crack. Duhamel states that he boiled a piece of wood, and then dried it upon a stove, but, in drying, it lost part of its substance, as well as the water contained; and upon a repetition, he found that it had lost still more of its weight. Four hours' exposure to steam or boiling water is sufficient for timbers of ordinary dimensions, and the drying afterwards goes on very rapidly, but it should be done as gradually as possible. The joiner frequently finds it necessary to steam or boil wood, to bend it into a particular curve; and also the ship-builder. It has been stated by writers on shipbuilding, that boiling increases the durability of timber; and in proof of this, they inform us that the planks in the bow of a ship, which are bent in this way, are never affected by the dry-rot.

It may now be inquired whether, after the most perfect seasoning, timber is secured against the process of decay? To this question a negative answer must be given. However well the timber may be seasoned, it will certainly rot if placed in a damp situation; the rapidity of the decomposition depending upon the nature and state of the wood, and the activity of the destroying agent. As the builder seldom attempts any other seasoning than that which depends upon drying his timbers, it is absolutely necessary that he should carefully avoid the rise of damp, and adopt every means in his power to prevent this evil. Timbers are usually placed in contact with walls, but it must not be supposed that this

F *

is sufficient to keep them from the access of damp, for
walls are frequently the conducting media. Brickwork
very readily absorbs moisture, and also throws it upwards, so that the ends of timbers are in contact with
the very source of mischief. To prevent the rise of
damp upwards, it is common to use, for a few feet above
the foundations, cement, a substance impervious to
water, instead of mortar, or to place between the courses
zinc or slate. But that these plans may be effective,
the basement walls should be surrounded with an open
area, for, if in contact with the earth on their sides, they
can be of no value. To prevent dampness from entering
in front, the brickwork should be covered with compo,
or some substance impermeable to water.

Another thing to be considered, for the security of
timbers, is to arrange, in every plan of a building, for a
perfect circulation of air. Ventilation is a most important requisite in the construction of a building, although
it is generally a matter of very little importance in the
consideration of those who have to plan or construct
buildings. The ventilation of roofs is by no means
difficult; but there are often so many obstacles to the
ventilation of flooring, that the designer will not give
sufficient attention to his subject to provide against
them. These things, however, are not matters of specu-
lation, to be attended to by those who have no higher
employment, but are absolutely necessary for the construction of a work that is intended to survive the
builder.

But we must pass from this subject, to a consideration of some of those plans which have been proposed,
to secure well-seasoned wood from the effects of dampness, and the ravages of insects; though it must be
confessed that but few of them have been successful.

Attempts have been made, from a very early period, to prevent the destruction of wood, by impregnating it with some substance capable of restraining its ravages. The muriate of soda, or common salt, has been thought a good preservative against decay, when the wood is thoroughly impregnated with it. The wooden posts which support the roof of a salt mine are said to be preserved by the constant infusion of salt, and that a vessel, covered with fungus, will have her timbers cleaned by immersion in salt water. Whatever may be the advantages of this process, it is quite certain that it can never be extensively employed; for the salt absorbs water so readily, that the timbers would be constantly damp.

In the year 1760, a Mr. Jackson proposed to immerse timber in a composition of muriate of soda, Epsom salts, lime, potash, salt water, and other substances; but neither he nor any body else could ever discover the value of this process. This person was permitted to prepare some timber to be used in the National yards, and it was found that vessels built with it were less durable than those in which unprepared wood was used.

Sulphate of iron, or green copperas, in water, has been recommended as a good mixture, in which to place wood that is to be used for the purposes of building. It is said that timber boiled in a solution of sulphate of iron, becomes so hard when dry, that moisture cannot penetrate it. This may possibly be the case, but the change must be effected by the removal of some portion of woody fibre, and the admission of the sulphate in its place; in the same manner as the wood found in the London clay has been fossilized by that substance.

Lime has been recommended as a preservative against the decay of timber. There is a difference of opinion among writers, as to the value of this substance for the particular purpose. It is well known that quick-lime, with moisture, rapidly destroys vegetable matter; but Mr. Tredgold says, that a large quantity of fresh quick-lime, in contact with wood, absorbs the water, hardens the sap, and thus, keeping it in a perfectly dry state, renders it very durable. This gentleman quotes the opinion of Mr. Chapman, who says, that vessels employed in the Sunderland line trade have been forty years old without needing any repair, or showing the slightest evidence of decay in the timbers. A writer, who recommends the impregnation of wood with lime, says, that wood buried in the earth, and surrounded by lime, is protected from the ordinary causes of decay. But Dr. Birkbeck objects to the plan, for he says, assuming such principle to be correct, there is a great inconsistency as to the effects produced upon animal and vegetable matter, and there can be no doubt that the substance which destroys the one, will destroy the other.

The attention of scientific men has been recently directed to the experiments made by Kyan; and from the very excellent exposition of his plan, by Dr. Birkbeck, we are induced to hope that it may be found highly advantageous. Having made a great number of experiments, with a view to ascertain the primary cause of vegetable decomposition, he was at last convinced that albumen was that cause, and that to neutralize its effects would be to prevent decomposition. Some plan was required similar to that adopted in tanning. The gelatine in animal bodies is quite as liable to decomposition as the albumen of vegetables; but when tanning, the infusion of oak bark is combined with it, the destructive properties are lost, and the animal matter becomes durable, and almost incapable of decay. Reasoning upon this effect, Mr. Kyan imagined that it might be possible to prevent vegetable decomposition, by causing the albumen to form a combination with some other substance; and, knowing the affinity of corrosive sublimate for the albumen, he entered upon a series of experiments, which led him to propose the use of that substance as a protection for timber.

A few extracts from the published lecture read by Dr. Birkbeck, before the Society of Arts, may put the subject more clearly before the reader.

" Mr. Kyan inferred that, as wood consists of various successive layers, in which the albumen, or juices containing albumen, circulated freely; it is quite certain that, as these juices within the wood, with the watery parts, fly off by the leaves, the albumen remains behind; and it is probable that this albumen, which from its nature is peculiarly prone to enter into new combinations, is the thing in wood which begins the tendency to decomposition, and produces ultimate decay; whether that decomposition is attended with the formation of cryptogamic substances, or whether in the less organized form, the change occurs with the simple production of what has been called the Dry Rot. He (Mr. Kyan) conceived, therefore, that if albumen made a part of wood, the latter would be protected by converting that albumen into a compound of protochloride of mercury and albumen; and he proceeded to immerse pieces of wood in this solution, and obtained the same result as that which he had ascertained with regard to the vegetable decoctions. Having done so, it became necessary to employ various modes of experiment, as well as comparative experiments. Now it is not clear in what part of the wood the vegetable albumen may be found, though it exists more especially in that part of the tree which is denominated the alburnum or sap, and is found between the heart wood and the innermost layer of bark. The experience of all practical men has confirmed the opinion, that this portion of wood is the first to decay.

"It is probable that, as the alburnum becomes successive layers of wood, it loses a quantity of albumen; or that, in consequence of the pressure which takes place by the addition of each successive layer, it becomes so situated, as to lose a part of its exposure to the vessels where a change may occur, and therefore becomes in some measure protected: for that which is one year alburnum or sap, may be, and indeed generally is, proper wood the next.

" The mode in which the application of the solution takes place is in tanks, which may be constructed of different dimensions, from twenty to eighty feet in length, six to ten in breath, and three to eight in depth. The timber to be prepared is placed in the tank, and secured by a cross-beam to prevent its rising to the surface. The wood being thus secured, the solution is then admitted from the cistern above, and for a time all remains perfectly still. In the course of ten or twelve hours, the water is thrown into great agitation by the effervescence, occasioned by the expulsion of the air fixed in the wood, by the force with which the fluid is drawn in by chemical affinity, and by the escape of that portion of the chlorine, or muriatic acid gas, which is disengaged during the process. In the course of twelve hours this commotion ceases, and in the space of seven to fourteen days, varying according to the diameter of the wood, the change is complete, so that as the corrosive sublimate is not an expensive article, the albumen may be converted into an indecomposable substance at a very moderate rate, and the seasoning will take place in the course of two or three weeks."

Mr. Kyan's method of seasoning has been already tested under circumstances so severe, that they may be said to have proved its efficiency. A piece of oak was five years in the fungus pit in Woolwich yard, a place notorious for the rapid and almost instantaneous destruction of vegetable matter, and it was as sound when taken out as when put in. This was the most severe test to which the method could be subjected, and its having sustained the trial is a proof of the value of the discovery. It has, however, been objected to the process, that the impregnation of timber with corrosive sublimate must unfit it for use in ship-building; but Mr. Kyan has furnished evidence to the contrary, and in our opinion proves that salubrity is one advantage. We strongly recommend the builder to make experiments himself upon wood prepared by Mr. Kyan, by using it in places where decay is rapid.

FRAMING OF TIMBERS.

When timbers are framed together, it is with the intention of supporting some weight, or resisting the strains to which the materials may be exposed, in the situations where they are to be placed. Horizontal or vertical timbers are not always of themselves sufficiently strong to sustain the pressure to which they may be subject, but they need assistance; and it then becomes a question, how can the materials intended to assist be best applied, and what are the smallest scantlings that can be adopted? Two things must be studied, stability and economy. It has been often stated that these two results cannot be accomplished by the same arrangement; but as the forces which are to be opposed have usually a direct application, so the system by which they are to be resisted may usually be of a simple construction. We have no doubt that those systems of framing are most effective, which are most simple, provided that the designer accurately determines the direction and intensity of the forces to be opposed, and judiciously applies the arrangements intended to resist their pressure. But this is not always done, partly from a want of knowledge in those who undertake public works, and partly from the insufficiency of those results which have been obtained by experiment. When we speak of the strength of timber, we shall have occasion to refer to the uncertain character of the principles upon which we depend in all our calculations, and, if it should be found that we have no means of accurately estimating the weight or pressure that a timber of given dimensions is capable of supporting, it must be then evident that, however accurate the means by which we estimate the forces, they are altogether inadequate to the deduction of proper results. But our first object is to explain the nature of the forces which are exerted by the several parts of a building, and the means by which they are to be resisted.

Composition and Resolution of Forces.

There are two great mechanical principles, which lie at the base of all proper attempts to estimate the nature of the forces which may be exerted upon substances in particular situations; these principles are called the composition and the resolution of forces.

The resolution of forces is the means of finding any two or more forces, which may resist or control the pressure of any one force. The composition of forces consists in finding the direction and amount of one force, that is capable of producing the same effect as two or more forces, acting in different directions. This is, in fact, only the reverse of the resolution of forces, and the two are, strictly speaking, but one principle; and if the one process be understood, the other must be almost so of necessity. Nor may the student pass over this part of the work, under a fear that it is too mathematical for him to understand; for he can never be certain that the roofs or other framing, which he may design, will support the weights they are intended to carry, if he does not know how to calculate the action of the weights or forces by which they may be pressed.

Let B D (*fig.* 1,) be the king-post of a roof, and let B A, B C, be the rafters: they are framed together for the purpose of carrying some weight; and the question is this, are they sufficiently strong to carry the weight which is to be placed upon them? To determine this we must refer to the resolution of forces. To put the problem in as simple a manner as possible, let us suppose some determined weight to rest upon the point B. Then, by some scale of equal parts, draw a line B, *d*, equal to the number of pounds, hundredweights, or tons, resting upon the point B, and draw *d a* parallel to B C, and *d c* parallel to B A. Now measure the line *a* B by the same scale, and it will give the number of pounds, hundredweights, or tons, by which A B is strained, and *c* B will give the strain upon B C. But, in the drawing affixed, the rafter B C is longer than the rafter B A; but this does not at all affect the weight, for it remains the same, whatever may be the length of the beam which carries it; but it is necessary to remember that, by increasing the length of the beam, it is rendered less capable of supporting the weight, and a proportionate increase of dimensions must be allowed. But should the direction of the beam be changed, a very different result will be

G

obtained, for in every case the pressure will be increased or decreased. The strain upon the beam B A (*fig.* 2) will now be measured by the line *a b*, and that upon B C by *b c*. In fact, a very slight alteration of position may, under certain circumstances, enormously increase or decrease a strain. It will be scarcely necessary to explain how two or more forces may be composed, and the single force, acting in a certain direction, be calculated.

Leaving the subject of the composition and resolution of forces, after a statement of the principle, we may proceed to explain the construction and arrangement of those parts of a building which belong to the carpenter. And, first of all, we may speak of roofs.

The construction of Roofs.

The roof of a building is that part which is intended to protect the interior from atmospheric changes, and at the same time to tie and strengthen the fabric itself. But in carpentry the term has a much less extensive meaning, and signifies the timber-work which is intend-ed to support the covering. The construction of roofs should always be regulated by scientific principles, for it is not only necessary to prevent it from straining the walls on which it rests, but it should strengthen them. Builders generally err in making roofs too heavy, which is a great fault, as the s ability of the building is im-paired, and a useless expense is incurred. There are many ways of constructing a roof, and they are not all equally suitable to the same situation. The span, the weight to be carried, and the country in which it is erected, should all be considered. The simplest method of constructing a roof is to place horizontal timbers from wall to wall; but this method is only suited to very short bearings, and does not readily throw off the water which may fall upon its covering. The Egyptians constructed flat roofs. To prevent this inconvenience, a roof may be made as an inclined plane; and such a construction has advantages, though its want of uniformity and beauty, and also its want of strength, proportioned to the amount of timber employed, are objections to its use; but still it is stronger than the flat roof, and readily carries off the water that may fall upon it. The best form for a roof is that in which there are two sides, equally inclined to the horizon, and resting in a line called the ridge of the roof. The angle which the inclined side forms with the horizon, is called the pitch. In countries where there is a cold climate, and snow is apt to fall in large quantities, the roof is high; in warm countries, the roof is low. The Greeks generally made their roof so as to have an inclination of from twelve to fifteen degrees; the Romans made theirs higher. In Gothic architecture the roof is generally high-pitched, and it is so consonant with the style, that it often forms a prominent feature in these buildings. There are not so many advantages in high pitched roofs as most persons suppose, and there are many disadvantages. The additional force of the wind upon a high roof is a serious objection; and when parapets are employed, it is so far from preventing the effects of a heavy fall of rain or snow, that the gutters are so filled that the pipes cannot carry off the water fast enough, or, being stopped by the dirt carried down by the velocity of the water, an overflow is occasioned. The height of roofs is now generally between one-third and one-sixth of the span.*

* The term *true pitch* in England generally means that the two sides of the roof measure breadth and half of the building.

It is the carpenter's business to frame the timbers of roofs, and sometimes he is required to design them; he should therefore know how to obtain the strength and other qualities required, with the smallest possible amount of timber.

A piece of timber, in whatever way it may be placed, except when vertical, will bend or sag, that is to say, its upper side will be formed into a concave surface. The more horizontal the timber is placed, the more it will always sag, and as the distance between the points on which it rests is increased, so it has greater liabilities to bending. To prevent this effect as much as possible, arrangements must be made for the support of the beam in some intermediate points. Now, it may be supported from either above or below. If there should be any walls between those on which the ends of the timber rest, these will be sufficient for all the purposes required; if not, the same result must be produced by a system of framing.

The timbers which compose a roof are known by different names, according to the uses for which they are employed, and the situations in which they are placed. The principal timbers of a roof are the following, but they are not all used in every roof: the tie-beams, wall-plates, collar-beams, king-posts, queen-posts, struts, principal rafters, common rafters, ridge-piece, collar-beams, purlins, and pole-plates.

The TIE-BEAM (A,) (*fig.* 3,) is a horizontal piece of timber, which extends from wall to wall, and rests upon the WALL-PLATES (B) at each end. It is employed for the purpose of connecting the feet of the principal rafters (c,) which would otherwise have a tendency to push out the walls, by their own weight and the weight of the materials placed upon them. In roofs of large span, it is necessary that the tie beam should be well supported in some point or points, between the ends on which it is supported, for if this be not done, it will sag and draw either one or both of the principal rafters towards its centre, and thus destroy the stability of the framing. The KING-POST (D) is sometimes used for this purpose. It is a piece of timber placed in a vertical position, connecting the point where the two principal rafters meet, and the centre of the tie-beam.

When the king-post is not thought to be sufficient to support the pressure which may be on the framing, QUEEN-POSTS (B) (*fig*. 4,) may be used, which are pieces of timber placed in an upright position, supporting severally the two rafters, and equidistant from the centre of the truss. The horizontal piece of timber (c,) which connects the heads of the queen-posts, is called a straining-beam; and that which connects their base, so as to prevent the struts from pushing them nearer to each other, is called a straining-sill. Those pieces which are placed in pairs, to assist in supporting the principal rafters, are called struts; they are frequently used to unite the rafters and the base of the king-post. Any horizontal timber above the tie-beam, is called a collar-beam. The ridge-piece (H) is that piece of timber which forms the apex of the roof, and is supported by the heads of the principal rafters or the king-posts, and in its turn supports one end of the common rafters. A pole-plate is a beam over the walls, supported by the principal rafters or the tie-beam, and is intended to carry the lower ends of the common rafters. Purlins (E) are horizontal timbers, between the pole-plates and ridge-piece. The small spars (c c) which are parallel to the principal rafters, and are supported by the ridge-plate, purlins, and pole-plates, are called common rafters.

G 2

The Dimensions of Timbers used in a Roof.

However accurately a roof may be designed, it is unfit for its purpose if the dimensions of the parts be not accurately proportioned. To accomplish this, some experience is required, and a knowledge of the strength of timbers, under particular circumstances. Some authors have given rules for the finding of these dimensions, but, although they have undoubtedly some value, many experiments must be made, before we shall be in possession of those data which will warrant the general application of the rules.

At present, the designing of roofs is governed almost entirely by experience, and no fixed laws can be appealed to. There are two things to be secured, a sufficient strength to support the weights to be carried, without sagging, and to do that without burdening the walls or other parts of the building over which the roof is thrown. This is not always an easy task, for roofs are sometimes to be made in such forms, as prevent the adoption of those means which would otherwise immediately accomplish the object. Sometimes a very large roof must be made flat, at other times a lantern-light must be provided in its centre; and, in a third case, it may be necessary to erect a dome. Now in designing for these and other roofs, attention should be paid to the character and success of similar works already executed; and the artist should study the points of similarity and difference between these and his own work, so as to provide against dangers which may peculiarly affect his building.

Examples of Roofs.

Fig. 5 is a roof, the rafters of which are only supported by a collar-beam, (c), which acts in part as a tie;
but this arrangement is so feeble, that it should never
be used over a space where the span is more than fifteen
feet.

In *fig.* 6, there is the addition of a tie beam, (A), and
the strain is here thrown from the collar to the tie-
beam; the former being compressed, the latter in a
state of tension. As there is no arrangement in this
truss to support the tie-beam, and to prevent it from
sagging, it is unfit for a span of more than twenty-five
feet.

To prevent the inconveniences resulting from the
sagging of the tie-beam, a king-post, (P), and struts,
(s s), may be introduced, as shewn in *fig.* 7. This
form of roof is very well adapted for a span of twenty-
five feet.

For a span of thirty to five and forty feet, the truss
represented in *fig.* 8 is very well suited, and is now very
commonly adopted by architects and builders.

Floors.

The timbers which support the flooring-boards, and
the ceiling of a room beneath, are called, in carpentry,
the naked flooring.

There are three kinds of naked flooring: single, double, and framed.

Single flooring is that in which there is but one series
of joists, as shewn in *fig.* 9, where A A A are joists, and B

the flooring boards. To make a single floor as strong as possible, the joists should be thin but deep, sufficient thickness being always allowed for the nailing of the flooring-boards. Two inches by six is the smallest dimension for joists; for a length of twenty feet, they should be about three inches thick, and twelve inches deep.

Sometimes the joists cannot have, in a particular place, a bearing upon the walls, and then a piece of timber is framed between the nearest joists. This is done where flues, fire-places, and stairs, interfere. The timber thus used is called a trimmer, and the two joists on which it is supported are called trimming-joists, and should be made a little stronger than the common joists. Thus, in fig. 10, A A, are common joists, B B, trimming-joists, and C, a trimmer. When the bearing is more than seven or eight feet, the joists should be strutted; that is to say, short pieces of board should be fitted between the joists, so as to form a continued line from wall to wall. These struts greatly strengthen the floor, and prevent the joists from sinking; but it is not desirable to mortice them into the joists, as that process has the effect of weakening the joists themselves.

Double flooring is that in which there are two tiers of joists, the binding joists, as A A, in fig. 11, which in fact support the floor, and the bridging joists B B. In this kind of flooring, the binders extend from wall to wall, and the bridging joists are notched down upon them. Beneath the binders we have a third tier of timbers (D), which are pulley morticed into the binders, and are called ceiling-joists.

When the binding joists are framed into a large piece of timber, called a girder, the floor is said to be a double framed floor. Thus in fig. 12, A is the girder, B, a binding-joist, c, a bridging-joist, D D, ceiling-joists, and E, flooring-boards. This kind of floor is decidedly the best when it is necessary to provide for a good and even ceiling; for, although single floors may be made very strong for a great bearing, yet the ceilings are always liable to crack.

It is not easy to obtain timber for girders, of much more than twenty feet scantling, and they are therefore trussed. Trusses are used in both floors and roofs, but we have not thought it desirable to interrupt the course of explanation we have given, by a reference to any particulars concerning this branch of carpenter's work, yet it is necessary that we should now make a few remarks upon it.

TRUSSES.

When timbers are so framed together as to support weights, they are called trusses. Now, it frequently happens that a piece of timber, in itself incapable of supporting a weight, may, when cut into scantlings of different dimensions, and framed together, not only carry that weight, but also support a much greater. The bow and string roof, invented by Mr. Smart, is an example in point. Let A A, in *fig.* 13, be a piece of timber, which we will suppose to be insufficient of itself to carry a particular weight, from this cut the pieces *o, s, c, b,* and *o, s, d, c.* Then let these pieces be raised as in *fig.* 14, and a key be placed between them at the apex; and it will form a very strong truss, which may be made still more capable of resisting a strain, by the application of struts.

The principal rafters of a roof are so called because they are trussed. It is not necessary to truss all the

rafters in a roof, and it would be very expensive to do
so; and therefore, trusses are placed at particular distances from each other, according to the weight to be
carried, and they are formed in different ways, according to the span over which they are thrown. The
planning of these is one of the most difficult tasks to be
given to the student; and to design them successfully,
so as to avoid a waste of timber and secure an adequate
strength, requires, in the first place, an accurate knowledge of principles, and in the second, a careful study of
the combinations which have been employed, under
particular circumstances, by professional men.

It has been already stated, that girders are sometimes
trussed, and should always be, when their bearing is
much more than twenty feet. Writers have differed as
to the value of the different methods of trussing girders,
and practical men are by no means agreed as to the
best forms and arrangements to be adopted. We have
often seen trusses which, so far from strengthening the
girders, have decidedly weakened them. Large girders
are sometimes sawn down the middle, and when reversed, are bolted together with slips of wood between
them. It has been supposed that this strengthens them,
and is adopted for this purpose; but the supposition is
erroneous, though the plan is certainly a good one, for
it allows a free circulation of air between the pieces,
and facilitates the emission of any dampness that may be
in the timber.

For many years carpenters were accustomed to truss
with oak, but, by experience, they discovered the impro-
priety of that plan, though indeed it might have been
suggested by the fact that oak is not much less susceptible of compression than fir.

A strong girder may, in fact, be made as strong as

any truss of the same depth, by bolting two pieces of timber together, or by confining them with iron hoops; the ends of the girder being smaller than the centre, so as to allow the hoops to be driven tighter, and confine the beams.

In *fig*. 15, we have given a representation of a strong truss girder, the truss-post and the abutment pieces being made of wrought iron.

OF CONNECTING TIMBERS.

It is sometimes impossible to obtain timbers of the length required for the several parts of a building, and it is then necessary to join two or more pieces together, so as to form them into one piece, and to injure the stability as little as possible. This process is called scarfing, and the parts of the joints which come in contact are called scarfs, and are usually connected by iron bolts.

There are many ways of scarfing, every builder adopting that one which appears to him the best under the circumstances in which the timber is to be employed. Two or three different methods may be mentioned, leaving the student to examine those which he may happen to meet with in practice, and the various designs which have been given by writers on the art of building.

Fig. 16 shows the means of scarfing without diminishing the length of the pieces. This is done by the introduction of a third piece, having the form of steps, and all the pieces being united together by bolts and plates.

Fig. 17 is a representation of a scarfing, which is very simple, and frequently used, though there is a considerable loss of timber. The pieces to be united are connected by iron bolts, an iron plate being placed on both sides.

Fig. 18 represents a form of scarfing, adapted to a beam, which has to support a cross strain. In many arrangements, the whole strain is supported by the straps and bolts, but in this they do not, in consequence of the indentation.

There are many complicated modes of scarfing, but these are mostly to be avoided. There are some persons who judge of the merits of designs, by the intricacy of the labour; but it usually happens that simplicity is *an element of strength,* as well as of beauty. If this principle were commonly acknowledged, much valuable time, talent, and capital, would be saved; and men, engaged in designing the workmanship for buildings, would learn the necessity of being regulated by established laws. This is particularly evident in relation to scarfing, for, as its strength altogether depends upon the accuracy with which the indents are formed, nothing can be more absurd than to make them in so complicated a manner as to prevent the necessary accuracy of execution.

TIMBER PARTITIONS.

Rooms and passages are often separated by timber partitions, which are so formed as to be covered with lath and plaster. In *fig.* 19 we have given a design for the framing of a partition, with a door through it; A A are the door-posts, B the head, c the sill, D D are braces which support the quartering, and are assisted by the struts, E E. It will be quite evident from a glance at the drawing, that the door-posts help to sustain the braces and struts; while they in return prevent the fall of the door-posts. Braces may be introduced in various ways, but strength is the object for which they ought to be introduced, a circumstance which is very frequently entirely forgotten by carpenters. It is strange that in so simple and easy a matter men frequently err, and waste both time and material in a provision for strength, where it is not required; while, at other times, labour and material are lost by neglecting to provide sufficient strength. In some instances it may be found desirable to introduce a simple truss into a design for partitions.

JOINTS.

It is frequently difficult, in an explanation of the processes adopted by those who work in wood, to separate the duties of the one from the other. The carpenter has little to do with joints, for they always weaken a combination of timbers, though they cannot be avoided; and his principal objects are strength and economy. We shall not, therefore, attempt in this place to explain the several kinds of joints employed in the union of woodwork, but leave this branch of our subject till we consider the joiner's work in particular.

The carpenter usually connects his timbers either by notching, or by mortice and tenon. Dovetail joints are sometimes used in carpentry, but they ought not ever to be adopted, for they will always draw when the timber shrinks, and the oblique surface of the dovetail tends to force the timbers apart, acting as though it were a wedge.

We have now taken a very general sketch of the work to be performed by the carpenter, and the manner

H

in which it ought to be done. In every portion of building, an acquaintance with the principles of some branch of science is necessary, but no workman requires so extensive a knowledge of scientific truths as the carpenter. Nor must he be satisfied with the casual examination of the influence exerted by bodies, as acted upon by the laws of gravity, and the capabilities of resistance possessed by particular substances, but he must trace the many complicated circumstances under which their several effects are modified. A knowledge of principles must precede the application of an acquaintance with detail, but the details are useless without principles.

THE JOINER.

THE business of the joiner is distinct from that of the carpenter, for it has relation to the more ornamental parts of the art of building;—the construction of wood-work is designed to please the eye, rather than to add to the stability of an edifice; whereas, the carpenter is concerned more with the solidity and stability, than the beauty or decoration. It will therefore be necessary that some remarks should be made, calculated to assist the student in the attainment of this branch of the art of working in wood, that beauty and solidity may be united. Nor must it be supposed that, in the introduction of facts already well known to some of our readers, we are writing that which everybody knows; for, in the preparation of this volume, we have aimed to render ourselves intelligible to those who are least acquainted with the subjects we explain.

WOODS.

White and yellow deal, wainscot, or American oak, and mahogany, are more frequently used by the joiner than any other woods. Having already spoken of these in the former chapter, it will be unnecessary that we should now do more than mention some of their qualities, and the uses to which they are applied.

Deal is generally imported into this country in lengths of from six to fourteen feet, though the most common length is twelve feet, being then generally cut to the greatest advantage. The blocks are about three inches thick, and nine inches wide. In choosing deals, those must be selected which are most free from knots and shakes, and appear to have the closest grain, the coarser ones being reserved for common work. The yellow deals are in general straighter in grain, and have a smaller number of knots than the white; this wood should therefore be chosen for the styles of doors, and for framing; and it works clean and sharp for the striking of all sorts of mouldings.

Wainscot is imported in logs of different dimensions. Such logs should be chosen, as appear to have been cut from a straight tree and have a clear grain, free from streaks of a lighter colour, for such parts are of softer texture; they are, in fact, but the beginnings of decay in the timber, and are termed by workmen doughty parts of the wood. If a log of wainscot be cut in the same direction as the beat of the wood, the boards will be variegated, and have a handsome grained appearance; but, if it be cut contrary to the direction of the beat, it will have a uniform appearance: the first is proper for panels, and places where it is intended to expose a large surface, and the latter for the styles and frames of doors, sashes, &c., as it mortices better and is less likely to split.

Mahogany is only used for the best kind of work; for doors, sash-frames, and baluster-rails. There are many kinds of mahogany, but they are classed by workmen under two general terms, Spanish and Honduras. The Spanish is considerably harder than the Honduras; and has not so commonly a variegated grain, though it is very handsome when a fine grained piece can be obtained. The Honduras has a spongy texture, and is often very cross-grained, which contributes to its mottled and variegated appearance; it is then extremely difficult to bring it to a smooth face. When the Honduras is good, it is very proper for the panels of doors; but the Spanish should always be used for framing, for mouldings, and for hand-rails, as it is less liable to break when cut out on the sweep; its texture being strong and its grain even. There is another kind of mahogany, known by the name of Ratteen, which is often employed for panels, as its dimensions are large enough to prevent jointing; and, as it is of a reasonable price, it is sometimes substituted for deal, particularly for curvilinear work that is to be painted. It may be used with advantage for the facia of shop-fronts, sign-boards, and other works liable to the changes of wet and dry, and exposed to the sun and air; for, however well the joints may be formed by the workman, they are always liable to fly when used in these situations.

Glue.

Before we speak of the methods employed in joining woodwork, it will be necessary to make some allusion to a substance, glue, which is in constant use by the workman; and, as its quality is of great importance to the joiner, we must speak of the tests by which to ascertain its adhesive properties, that the workman may know how to select that which is best, and to reject that which has not the requisite adhesive property. Glue is made from the skins and sinewy parts of animals, or from the skins and cartilage of fishes. The glue that is made from land-animal substances is considered to be better than that made from fishes; though isinglass, which is made from the air-bladders of a large fish found in the Russian seas, is one of the strongest with which we are

н*

acquainted; but its price in the market prevents the
joiner from employing it. From the chemical experiments which have been made, it appears that the glue
manufactured from the skins of animals, is superior to
that which is made from the sinewy or horny parts, as
well as that which is made from the skins of fishes, not
being so readily affected by the moisture of the atmosphere. The workman, therefore, always prefers animal-
glue to that which is called fish-glue, though the latter
is often sold as glue of the best quality. Some directions may be given to enable the joiner to choose this
necessary cement, and to judge of its adhesive qualities.
All glue, in the cake, is subject to the effects of dryness
or moisture, which, in the atmosphere, are constantly
changing, becoming soft in damp weather, and brittle
in dry. But the different kinds are differently affected.
Glue should be purchased in dry weather, for that which
is then soft is not of so good a quality as that which is
crisp. Some opinion of the quality may be formed by
its transparency, for that which is the most transparent
is the best. If it be possible to make an experiment
with a sample of the article, before a quantity is purchased, a cake may be immersed in water, in which it
should remain two or three days, and, if the glue be
good, it will not dissolve, but swell; but if it be of inferior quality, it will partly, if not wholly, dissolve in
water; from which it follows, that that which is least
dissolved in cold water is the best, or possesses superior
qualities of adhesion, and will be least affected by moisture or damp. Another test is, that being dissolved in
water by heat, the glue is best which is most cohesive,
or which may be drawn into the thinnest filaments, and
does not drop from the glue-brush like water or oil, but
extends itself in threads when falling from the brush or

stick ; and this it will always do if the glue possess the requisite properties.   These tests will enable even the inexperienced workman to judge of the quality of the material offered to him for sale ; and, in a very short time, he will find no difficulty in selecting that which will give firmness and solidity to his work.   It may be worthy of remark, that the glue made from the skin of old animals is much stronger than that of young ones.

Glueing Joints.

In general, nothing more is necessary to glue a joint, after the joint is made perfectly straight, or, in technical terms, out of winding, than to glue both edges while the glue is quite hot, and rub them lengthways until it has nearly set.   When the wood is spongy, or sucks up the glue, another method must be adopted, one which strengthens the joint, while it does away with the necessity of using the glue too thick, which should always be avoided; for the less glue there is in contact with the joints, provided they touch, the better; and when the glue is thick, it chills quickly, and cannot be well rubbed out from between the joints.   The method to which we refer is, to rub the joints on the edge with a piece of soft chalk, and, wiping it so as to take off any lumps, glue it in the usual manner; and it will be found, when the wood is porous, to hold much faster than if used without chalking.

To make a very strong Glue.

An ounce of the best isinglass may be dissolved, by the application of a moderate heat, in a pint of water.

Take this solution and strain it through a piece of cloth, and add to it a proportionate quantity of the best glue, which has been previously soaked for about four and twenty hours, and a gill of vinegar. After the whole of the materials have been brought into a solution, let it once boil up, and strain off the impurities. This glue is well adapted for any work that requires particular strength, and where the joints themselves do not contribute towards the combination of the work; or in small fillets and mouldings, and carved patera that are held on the surface by the glue.

Of the different Methods of joining Woodwork.

Every workman, or rather every person who has seen any thing of carpenter's work, must be acquainted with the meaning of the terms dove-tail, mortice and tenon, and grooving; but the knowledge of the meaning of the terms does not imply an acquaintance with the best methods of performing the several operations, which can only be obtained by practice. There are many workmen who are not aware of the proportion which a piece, made to fit into another, should have towards that into which it is fitted, so as to produce the greatest strength with the least possible waste of material; or how to proportion a joint, so that it shall not fail or give way before another. We shall, therefore, endeavour to lay down some rules, and produce some examples, which will be at least an attempt to bring into a general arrangement the principles of joining. In too many instances, the method of joining woodwork is regulated by no other rule than the fancy of the workman. It is not difficult to explain why joiners' work so frequently fails; why the parts separate with a trifling strain; or, from being bound too tightly together, fly and split in all directions. It is not so frequently from the bad execution of the work, as from the want of an adequate estimate of the strength required to resist the stress on the joint. We may then, for the use of the student, describe the several kinds of joints, or the methods of framing and joining timber; and, under each head, give such directions, founded on the principles of mechanics, as will enable the workman to proceed with some degree of certainty; and not, as is too frequently the case with artisans, observe no other rules than those which custom has authorized, and practice has made familiar.

Dove-tailing.

We have given, in the plates, several examples of dove-tailing. The parts which fit into each other are known by different names; the projecting piece, represented in fig. 20, is called the pin of the dove-tail; and the aperture into which it is fitted, as shewn in *fig.* 21, is called the socket. Now the strength of a dove-tail depends upon so proportioning the pin and the socket as to enable them to support, rather than destroy, each other. Let A B C D, *fig.* 20, be a scantling, which is required to be joined to another, by means of a single dove-tail. The strength of the joint depends on the form of the dove-tail, as well as on the proportion it bears to the parts cut away; we shall endeavour to lay down the principle on which the greatest strength may be secured. Having squared the end of the scantling, and guaged it to the required thickness, A I K L M, divide I M into three equal parts, at K and L. Let K L be the small end of the dove-tail, and make the angles I K G and M L H equal to about 75 and 80 degrees, respectively; and make

G E and H F parallel to A N and B O. Then cut away
the parts A I K G E N, and B M L H F O, and having
formed the socket to correspond, by marking the form
of the dove-tail on the top of the piece A B C D, *fig.* 21,
and cutting away accordingly, the pieces may be fitted
together, as shown in *fig.* 22. It may be here observed,
that the bevil of the dove-tail, that is, the angle I K G,
fig. 20, may be either more or less than has been mentioned, according to the texture of the wood. Hard,
close-grained woods, not apt to rive or split, will admit
of a greater bevil than those which are soft, or subject
to split; thus the bevil of a dove-tail in deal must be
less than in hard oak, or in mahogany. It is a great
fault to make a dove-tail too beviling, for instead of
adding to the strength of the joint, as some persons
suppose, it weakens it; for provided the bevil is sufficient
to prevent the possibility of pulling the pieces apart, the
less the bevil that is given the better. It must have
been observed, that there is a great difference between
the dovetail made by the cabinet-maker, and that by the'
joiner; the former has very little bevil, the latter very
much; the former looks neat, and is at the same time
strong; while the latter, appearing to aim at strength,
looks clumsy, and is, at the same time, much the weaker
of the two.

Fig. 23 represents the dove-tail in common use for
drawer-fronts. When it is required to hide the appearance of the joint in front, the board A B C D is cut with
the pin, and A E F B with the socket. The pins in this
sort of dove-tail are in general from about three quarters
of an inch to an inch apart, according to the size of the
pieces to be joined.

Fig. 24 represents the pin part of a lap dove-tail,
which, when put together, shows only a joint, as if the

pieces were rebated together, as shown in *fig.* 25. A B C D represents the pin, E F G H the socket; when put together, the line H G is only seen as a joint, and if the corner A B is rounded to the point G H, it will appear as if only mitred together. This kind of dovetail is very useful for many purposes where neatness is required, such as in making boxes.

Fig. 26 represents a still neater dove-tail; and, as the edges are mitred together, it is termed a mitred dove-tail; it is the same as that shown in *fig* 6, except that instead of the square shoulder, or rebate, in A B, it is cut into a mitre, and the other piece is made to correspond.

Another very neat as well as expeditious method of joining pieces of wood, and which is somewhat analogous to dove-tailing, is shown in *fig.* 27.· The joint is first formed into a mitre, and the pieces are then keyed together, either by making a saw kerf in a slanting direction, as at A B, or by cutting out a piece, as at C D, in the form of a dove-tail. The first method, A B, is called, amongst workmen, keying together; the second, C D, key-dove-tailing.

The last method to be mentioned is that shown in *fig.* 28, and may be termed mitre-dove-tail grooving; the part A B, being formed with shoulders cut to the required bevil, and a piece left for the pin dove-tail, which is inserted into the socket dove-tail, made to correspond to it in the piece C D, which has been previously formed into a mitre. This method, though not much employed, may be used with great advantage in many instances, particularly when it is required to join pieces together the lengthway of the grain.

Mortice and Tenon.

Under this head we shall endeavour to give some rules, necessary to be observed in attempting to proportion the parts of the mortice and tenon, so that they may be equally strong, or that the tenon may not be more likely to give way than the cheeks of the mortice; for this is the principal thing to be avoided. The workman frequently allows too little substance for the tenon, lest he should weaken the mortice, and sometimes he falls into the opposite error; facts which clearly prove, that he is not acquainted with a means of obtaining a maximum of strength with a given quantity of material.

Figs 29 and 30 represent a simple mortice and tenon. The dotted lines show the parts to be cut away. To shew the thickness of the tenon, and, consequently, the width of the mortice, we have here one tenon and two shoulders; that is, three parts, one of which is to be allowed for the tenon, and two for the shoulders; and this will in general be found the best proportion, for if the tenon be more than that, it will weaken the shoulders of the mortice. Now if we have, as is frequently the case, two tenons in one piece, as represented in *fig.* 31, there will be five parts, two tenons and three shoulders; so that each tenon will be one fifth of the thickness of the stuff, for the shoulders are all equal to the tenons. This rule may be generally observed, unless the tenon is at a considerable distance from the end of the stuff, and then something more may be allowed for its thickness, as the mortice is then not so liable to split; but it should in no case, however sound the timber, or tough the material, be more than two out of four parts; that is to say, it can never be safe to make the tenon more than half the thickness of the stuff, and that only under particular circumstances, as when the mortice is near the middle of the scantling, for the piece in which the mortice is cut would, in other cases, be considerably weakened.

There is frequently, in joiner's work, a shoulder at the bottom of the tenon, that fits into the piece in which the mortice is cut, as represented in *fig*. 32; and the tenon is divided into two parts, as there shewn, which, when the stuff is wide, is a good method, as it strengthens the piece in which the mortice is cut, without weakening, in the same proportion, the mortice itself; and we would suggest, in this case, that the piece B C, cut out from between the tenons A B and D C, be nearly, if not quite, one third of the distance A D; for if much less, the piece left within the mortice will add but very little to the strength of the piece in which the mortice is made; and the tenon would be stronger in proportion to the mortice-piece than necessary. It may be here observed, that if the width of the tenon be much more than four times its thickness, additional strength will be gained by dividing the tenons into two or more parts, as shewn in the figure, particularly if we allow a small piece at the bottom of the tenon, as represented in the drawing.

Grooving and Lapping.

It is not necessary to say much concerning this method of joining wood-work, it being analogous to that of morticing and tenoning. We shall, therefore, simply state, that when it is required to join two boards together, by means of a tongue and groove, the groove should never exceed one third of the thickness; and often, if the piece for the tongue be formed of hard wood and liable to split, one quarter of the thickness will be sufficient.

I

When a panel is let into a groove in the style, the joiner is often guided by the thickness of the panel itself, which should never be less than one-third the thickness of the style.

In making a groove across the grain, as for partitions, it will be best, in most cases, to make it about a fifth or sixth of the substance of the stuff. But, if the groove be formed into a dove-tail, one quarter of the thickness will be better, and the dove-tail should be made a little tapering, but not too much; it should, in fact, be so made as to go almost home, without requiring a blow from a hammer or mallet to drive it into its place, until it has nearly attained it; and all joints should be easily separated by a gentle blow before they are glued. In a lap-joint, that is, in lapping two pieces together, supposing them of equal thickness, half the substance of each should be cut away; and if of unequal thickness, the lap should be made in the thinner piece, of about two thirds or three quarters of its thickness, according to the substance of the thicker piece; thus endeavouring in this, as in all other cases, to avoid the weakening of one piece more than another.

Bending and Glueing up.

In bending and glueing up stuff for sweep-work, much judgment is necessary, and as the methods are various, we shall mention a few, which the workman may apply as occasion may require, one method being preferable to another, according to the nature of the work in hand.

The first, and most simple method is that of sawing kerfs or notches on one side of the board, thereby giving it liberty to bend in that direction; but this method, though very ready and useful for many purposes, weakens the work, and may cause it to break when strains are thrown on the piece. But a tolerably strong sweep may be made in this manner, if, after sawing the kerfs, (particular care being taken to make them regular and even, and to saw them at regular depths,) some strong glue be rubbed into each kerf. When bent into the required sweep, a piece of strong canvass should be glued over the kerfs themselves, and the glue be left to harden in the position to which the stuff is bent.

Another method is to glue up the stuff in thin thicknesses, in a cawl or mould, made with two pieces of thick wood cut into the required sweep. This method, if done with care, that is, making the several pieces of equal thickness throughout, of wood free from knots, is, perhaps, the best that can be devised for strength and accuracy. It is also a practice sometimes to glue up a sweep in three thicknesses, making the middle piece the contrary way of the grain to the outside and inside pieces, which run lengthwise. This method, though frequently used for expedition, is much inferior to the above, as the different pieces cannot shrink together, and consequently the joint between them is apt to give way.

A solid piece, if not too thick, may be sometimes bent into the form required. If a piece of timber be well soaked upon the intended outside of the curve, it may be bent into position, and, if kept in that position till cold, will retain the curvature that is given to it.

The only other method of forming a curve, necessary for us to mention, is that of cutting out solid pieces to the required sweep, and glueing them upon one another till they have the thickness required; taking care that the joints are alternately in the centre of each piece below it, something in the manner of courses of bricks one above the other. In this case it will be necessary,

if the work be not painted, to veneer the whole with a
thin piece, after it has been thoroughly dried and planed
level, and then made somewhat rough with either a rasp
or a toothing-plane. But the joiner must adopt one plan
or another, according to circumstances.

Scribing.

Scribing is the operation by which a piece of wood-
work is made to fit against an irregular surface. Thus,
for instance, the plinth of a room is made to meet, or
correspond with, the unevenness of the floor. To determine the portion which is to be cut off from a partition, or any wood-work where a floor or ceiling is
irregular, it is only necessary to open the compasses to
a width equal to the greatest distance between the
plinth and the floor, and passing one leg over the uneven surface, the other leg will leave a mark on the
plinth. If the wood be cut away on that line, a surface
will be obtained which will make a good joint with the
floor or ceiling. But the chief use of the art of scribing
is to enable the joiner so to connect the moulding of
panels or cornices, that when placed together they shall
seem to form a regular mitre-joint. This method has
certainly one advantage over the common method of
mitreing, for, if the stuff should shrink, little or no alteration will be made in the appearance, but, under the
same circumstances, a mitre would open, and the joint
would be shown. The method adopted is this; cut one
piece of the moulding to the required mitre, and then,
instead of cutting the other to correspond with it, cut
away the parts of the first piece to the edge of the first
moulding, which will then fit to the other moulding,
and appear as a regular mitre.

Finishing of Joiner's work.

As joiner's work is generally intended to increase the beauty of a building, and as the appearance depends much upon the manner in which it is finished, we shall mention a few principles which must be attended to; for, however well the work may be executed, so far as regards the strength and accuracy of the several joints, if the finishing be disregarded, whether the wood be in-tended to have its natural appearance, or to be varnished or painted, the elegance required cannot be obtained, if the joiner does not properly finish his work. When a joiner works in wainscot, oak, or mahogany, his chief object must be to obtain a surface perfectly smooth and even. When the framing is glued together, the glue which oozes out, and may be spilt upon the work, must be allowed to remain a few minutes to chill; it may then be carefully scraped off with a chisel, and the parts which cannot be thus cleaned may be washed, with a sponge dipped in hot water and squeezed nearly dry. This not only saves trouble in operations which follow, but prevents staining, which is always produced when glue is suffered to remain till quite hard, particularly on wainscot, which turns black in every joint or place where the glue is suffered to remain. After this operation, which, though it may appear tedious to some workmen, will be found a saving of time, the work should remain till perfectly dry; and when the joints and other parts have been levelled with a smoothing plane, the whole surface may be passed under a smooth scraper, and finished with fine glass paper. It will be sometimes necessary, when the grain is particularly cross, to damp the entire surface with a sponge "to raise the grain," and then again to apply the glass paper. The work will then be ready for polishing with wax, or for oiling, or varnishing, and the good appearance of the work will be in proportion to the time and trouble expended in the process.

In cleaning deal, the same precautions must be taken for the removal of glue left upon the joints, or spilt upon the work, as already described. This being done, the work may be cleaned off with a piece of glass paper that has been rubbed with chalk, or, in some cases, with a piece of hearth-stone. The work is then ready for the painter, but as there are knots and other places where the turpentine contained in the wood is apt to ooze out, either with or without the increase of heat, and thus spoil the appearance of the finishing, those parts are done over with a composition, and the process is called priming. This is properly the painter's business, but it must sometimes be done by the joiner for the sake of saving his work. The composition used for this purpose is made with red-lead, size, and turpentine; to which is sometimes added a small quantity of linseed-oil. Priming has also the advantage of preventing the knots from being seen through the paint. Some workmen omit, in this composition, the oil and the turpentine, but the size of itself is apt to peel off, and does not thoroughly unite itself with the wood.

Another method of cleaning off deal is sometimes adopted. When the surface has been made quite smooth with the plane, it is rubbed with a piece of chalk, and the whole is cleaned with a piece of fine pumice-stone, as in the former process it was done with glass-paper; but if the grain should be still rough, the work may be damped with a sponge, and the operation repeated when dry.

As in finishing interior work it is now customary to imitate the graining of different kinds of wood, it is necessary that the joiner's work should be well finished, for if a good even surface be not provided, it will be impossible for the painter to produce the effect he desires. Every defect in the ground will, in fact, be more visible under a delicate graining, than when the surface is covered with successive coats of colour; but even in the latter case, work well prepared will not only look better, but the colour will not be so apt to chip and peel off, as when the surface is not properly levelled.

To make Glass or Sand-paper.

As the paper used in cleaning off wood-work is of great service to the joiner, it may be necessary to give an account of the manner in which it may be manufactured, should the workman be ever placed in circumstances where it cannot be purchased. Take any quantity of broken glass, (that with a greenish hue is the best,) and pound it in an iron mortar. Then take several sheets of paper, (fine cartridge is the best,) and cover them evenly with a thin coat of glue, and holding them to the fire, or placing them upon a hot piece of wood or plate of iron, sift the pounded glass over them. Let the several sheets remain till the glue is set, and shake off the superfluous powder, which will do again. Then hang up the papers to dry and harden. Paper made in this manner is much superior to that generally purchased at the shops, which chiefly consists of fine sand. To obtain different degrees of fineness, sieves of different degrees of fineness must be used.

As in cleaning wood-work, particularly deal and other soft woods, one process is sometimes found to answer better than another, we may describe the manner of manufacturing a stone-paper, which in some cases will be preferred to sand-paper, so as to produce a good face, and is less liable to scratch the work. Having prepared the paper as already described, take any quantity of powdered pumice-stone, and sift it over the paper through a sieve of moderate fineness. When the surface has hardened, repeat the process till a tolerably thick coat has been formed upon the paper, which, when dry, will be fit for use.

To Polish Wainscot and Mahogany.

A very good polish for wainscot may be made in the following manner. Take as much bees-wax as required, and placing it in a glazed earthen pan, add as much spirits of wine as will cover it, and let it dissolve without heat. Add either one ingredient as required to reduce it to the consistence of butter. When this mixture is well rubbed into the grain of the wood, and cleaned off with clean linen, it gives a good gloss to the work.

Another polish may be made in the following manner; take of the best linseed oil one quart, to which add half a pint of the best spirits of turpentine, and a piece of lime about the size of a cricket-ball, broken in pieces. Let the mixture simmer, on a stove covered over, for two or three hours, then strain it through a coarse cloth, and it may be kept for use or used immediately. It must be put on the work with a brush, and allowed to remain for about four-and-twenty hours, after which it should be rubbed off with a woollen cloth, and the work be finished with a clean piece of linen.

If the colour is to be heightened, as well as a polish

to be given, a varnish may be used, which is made in the following manner: take one quart of linseed-oil and half an ounce of litharge, and let them simmer for an hour or two, and afterwards strain off the compound. Then take about half a pint of spirits of turpentine, and add to it as much pounded turmeric as will be sufficient to give the colour required. When this has been strained off, it may be mixed with the oil, and used in the same manner as the one just described; but, if the process be repeated two or three times, a day or two intervening between the applications, the effect will be increased.

A good polish for mahogany may be made in the following manner: take of linseed-oil one quart, alkenett root one ounce, and rose-pink half an ounce; stir them well together, and place them near the fire to simmer gently for an hour or two, then strain off into a clean pan. Apply the polish with a brush, and let it remain for about half an hour. Then take of the finest red brick-dust, sifted through a cloth, and dust it over the work. A piece of woollen cloth should be used in polishing, the wood being rubbed in the direction of the grain. A clean linen cloth and saw-dust should be used in finishing.

In the remarks we have made upon the works performed by the carpenter and the joiner, we have attempted to confine our attention to those facts which seemed to be most important for the student or the young workman; but, at the same time, we have not satisfied ourselves with a mere statement, but have endeavoured to explain the principles upon which the practice is founded.

THE MASON.

THE business of the mason has been properly divided into two parts, one having a relation to the substantial work of a building, the other to the ornamental. We shall not divide our remarks upon the art of masonry into two sections, for the purpose of retaining this classification, but, by keeping this arrangement in mind, we shall perceive what is necessary to be said, and the relative importance of the several subjects which come under consideration. The mason must know something of the quality of the stone he employs, its fitness for the purposes he requires, and the dimensions necessary to give firmness and durability to his work, at the same time securing the greatest possible strength, with the least quantity of material. He must also know the various methods of placing the stones, so as to get a proper joining, and whether they are to be held together by cement, cramps, or otherwise. These subjects are equally important to all workmen engaged in this department of building; but the man who professes to execute ornamental work in stone must study how he can make all the parts of his work harmonize or bear a just proportion to each other, and must pay particular attention, especially in works directly under the eye, to the accuracy and finishing of the several mouldings and ornaments which he is called upon to execute. This branch of masonry is allied to statuary, in which the more costly materials, such as marble and porphyry, are used.

We shall commence our observations on the art of

working in stone, by an allusion to the nature of the materials themselves, and describe the different kinds of stone and marble used in building ; and then offer some hints with regard to the several purposes to which they are best adapted, and the manner in which they are worked.

The stones used in building may be properly divided into two kinds, those of a sandy or gritty texture, which are incapable of bearing a polish; and those which, their texture being more compact and hard, are capable in a greater or less degree of being wrought to a very smooth face, and polished. The former are generally called stones, and the latter marbles; the former are used for out-door work, where expense both of material and workmanship must be prevented as far as possible : the latter, for those ornaments and conveniences within doors, such as slabs and chimney-pieces, in which appearance is more considered than necessity.

Every country has its building-stone, but the suitability of that stone depends upon the geological series which happens to appear upon the surface. It is very important to every community, that there should be, near to the locality in which it is settled, some mineral substances adapted for the purpose of building. It is much to be regretted that, in the establishments of colonies, this has not been sufficiently attended to, and that the people are consequently denied, not only those elegancies which add so much to the comfort of civilized life, but also the necessary material for an enduring structure. In such a place, there can be nothing in the external character of a city, that has a claim to an existence that can be called duration, but a few short years must give an entirely new appearance to all the structures, and to the city itself. Such a state of things is very likely to repress those arts which ought to be encouraged, and make society at large indifferent to those comforts, which can only be obtained with great expense, and which are constantly in danger of being removed by casual circumstances, or the ordinary process of decay, to which the very sources of comfort are exposed.

England is most advantageously situated in a geological sense. There is no tract of country in the world, of equal extent, that has such vast reservoirs of coal and metal, and there is none that has so much and so good building stone. Our marbles may not be so fine as those of Italy, but our building stones are good, and may be obtained at a comparatively trifling expense. The great improvements which have been made in the construction of bricks, and their strength and efficiency, have tended to do away with the necessity for the use of stone. We have, however, many evidences of the great antiquity of the art of building in stone. There are in existence, not only remains, but entire buildings of very ancient date, built entirely of stone; many of these are objects of wonder, even at the present day, some from the elegance of their forms, and others from the immense size of the materials of which they are constructed. But as we have more to do, in this little volume, with that knowledge which may be applied to practical purposes by the reader himself, than with theoretical opinions or past practices, we must confine our remarks to those subjects which may be useful to the student.

The stones most commonly used in England for heavy masonry, are the Riegate-stone, Purbeck-stone, Free-stone, Portland-stone, and Granite.

Riegate, or fire-stone, is a freestone, capable, as its name imports, of withstanding the effects of fire, and is therefore used in all those parts of a building where it is exposed to its action, such as hearths, ovens, and stoves. It is chiefly obtained from Sussex.

Purbeck is a hard greyish stone, and is chiefly used for pavements. It is capable, from its very compact texture, of being wrought to a very smooth face, and will bear a slight polish.

There are several kinds of freestone, and they are obtained in different places. When first taken from the quarries, freestone is, in general, very soft, but by exposure to the atmosphere it becomes much harder. It may at first be easily cut with a common saw, and may be worked almost as easily as a piece of timber; but after exposure to the atmosphere for a few weeks, it becomes very hard. Bath stone is one of the best freestones obtained in this country, and is preferred to all others when it can be procured at a moderate price; for it has, in an eminent degree, the property of hardening by exposure to the air, and is not apt to chip and peel as many others are. It is a fine sandy grit of a whitish colour, and, from the ease with which it is worked, is well adapted for chimney-pieces, jambs for windows and doors, the dressings of windows, and for other external work.

Portland stone is somewhat similar to the Purbeck, but softer and whiter; it is raised in much larger blocks from the quarry, and is of very extensive use in building; it will not, however, stand the fire, but well endures the vicissitudes of the weather. It is, perhaps, the best common stone for building, having sufficient hardness, durability, and equality of texture, for every purpose in building; added to which, its comparative cheapness, and the large size in which the blocks are or may be raised, makes it vastly superior to the Purbeck.

K

Granite has of late been very much used in building, particularly where strength and durability are required. This stone has a very hard crystalline structure, and resists the usual methods of working. It is reduced to the form required, by pecking, with a kind of hammer, somewhat similar to a pickaxe. It is found in large quantities in many parts of the west of England, particularly in that district called Dartmoor, near Plymouth, though it is a prevailing rock throughout Cornwall. It also abounds in many parts of Scotland; that which is brought from Aberdeenshire is much esteemed. This stone is particularly valuable in those situations where there is much wear, as, for instance, in the steps of public buildings, the curb stones of pavements, the pavement or carriage-way of roads, and the piers of bridges. Waterloo and London bridges are almost wholly composed of granite, both of which will long remain as monuments of the skill and talent of the architects who designed and constructed them.

It is impossible to give any classification of the stones which, from their beauty, as well as their costliness, are adopted by architects as the proper ornaments of the interior of buildings. They are all known by the general term, marble; but although they may all be brought to a fine polish, having a great hardness and firmness of texture, they differ from each other in structure, and in colour, and are known by specific names; thus we have the Italian, Egyptian, and other marbles; the porphyry, statuary, and alabaster; but all possess common properties, though they differ in colour and in texture.

There are some defects in marbles, which diminish their beauty, and consequently their value, while at the same time they add to the difficulty of working them. When a marble has an excessive closeness of texture, which renders it hard to work and apt to splinter, such as the black marble of Namur, it is said to be rigid. Thready marble is that which is full of filaments, and may be compared to wood of a soft and cross grain; this defect renders it difficult to work or polish. Brittle marble is that which crumbles under the tool, such are the white Grecian and Pyrennean marbles. Terras marbles are those which have some places softer than others, or those in which the texture is not equal throughout its substance; such is the Languedoc marble. There are also two general defects common in marbles, and worthy to be mentioned, which, increasing the difficulty of cutting and polishing, are well known to workmen; one they call *nails*, which may be compared to the knots in wood, the other they call *emeril*, which is occasioned by a mixture of copper or some other metal in the substance of the marble; this defect is common to white marbles, and knots to all.

We may be here permitted to recommend to the attention of the reader, a very fine collection of the British marbles exhibited in the great room of the Society of Arts, among which will be found some equal in beauty to the finest Italian; and we do hope that the encouragement given, and the rewards offered to investigators, will be the means of bringing to our notice as fine quarries in the British dominions, as those in the Italian states.

Of the different kinds of Masonry.

Masonry, in the general acceptation of the term, is the art of cutting or squaring stones, to be applied to the purposes of building; or in a more limited sense, it is the art of joining stones together, with mortar, or otherwise.

The ancients enumerate seven different methods in which they arranged the stones of their buildings. Vitruvius thus classes them; three of hewn or squared stones, three of unhewn, and one a mixture of both methods.

1. Net masonry. This is represented in *fig.* 33 within the area D E F G, where the stones are squared and placed upon one of the angles, their joints thus forming a net-like appearance. This method, though very neat, is wanting in firmness and strength; for the oblique position of the stones, in regard to each other, gives them a tendency to separate, rather than to form a compact assemblage of parts that unite in supporting each other. Whenever this form of masonry is employed, it is, consequently, necessary to keep the work together by a border of stones, having some other arrangement, one that is not only capable of supporting itself, but of overcoming the resistance of the net-like form. This is shown in the same figure at A B C; and, where the network is merely a casing of stone to the brickwork of a wall, it will be found to answer tolerably well, and looks very neat.

2. Bound masonry is that represented in *fig.* 34, and is remarkably strong. The perpendicular joints, in each course, fall directly in the middle of the stones composing the course below and the course above it; and while it has every requisite of solidity, the joints have, at the same time, a regular and pleasing appearance.

3. Greek masonry is that represented in *fig.* 35, where every alternate stone, as shown at A D, E F, and B C, is made of the whole thickness of the wall, and serves to bind together the stones which compose the external and internal faces of the building. This may be called double binding, as, from the perpendicular joints being somewhat similarly situated to that in bound masonry, it has

also an additional binding, by extending to the courses above and below it, thus forming a compact and durable wall, which resists every effort to separate in any direction.

4. Masonry by equal courses. This method of uniting stones, is shown in *fig.* 36, and only differs from the bound masonry in its being composed of unhewn stones, or rather in being formed of stones that are not so accurately cut, nor the edges so perfectly squared; it being only necessary that the external face should be level, and the horizontal joints at equal distances from each other; care being, at the same time, taken, that the perpendiculars are so situated as to bind the courses above and below them.

5. Masonry by unequal courses. This is represented in *fig.* 37, and is, like the last, formed of unhewn stones, without any regularity as to their size; it being sufficient that each course is made to bind with the preceding, and the only regularity observed is in the joining, which is necessary at each course, the courses themselves being of unequal thickness, as shown at A B C D.

6. Masonry filled up in the middle, as shown in *fig.* 38, is formed of unhewn stones of unequal courses, and the middle, as at D, is filled up with stones, thrown in at random among the mortar.

7. Compound masonry is, as its name imports, a mixture of the other kinds. It is represented in *fig.* 39, where the external course A B, is formed of bound masonry, and the corresponding internal course is at some distance from it, but held to the former by means of iron cramps, as shown at *a*, *b*, *c*, *d*, *e*, *f*; the space between being filled in with small stones or flints, thrown into the mortar.

K *

The Methods of Joining Stone.

As the strength and durability of masonry depends as much on the method employed, and the care taken, in making all the joints to correspond accurately with each other, as in the quality of the material employed, some remarks will be required in explanation of the methods of joining stone. We shall. therefore, enumerate the several means adopted by workmen, and, where necessary, notice the purposes to which each method is best adapted; giving some cautions to secure success in practice, and to save the workman unnecessary labour and trouble.

The joints in masonry are either secured by means of mortar, cement, or plaster of Paris, or the courses are held together by cramps, joggles, mortice and tenoning, or dove-tailing.

1. Joining by mortar, or by cements. It is absolutely necessary that the joints should be perfectly smooth, and touch in every part, and the stones must be so square as to bed well on each other, that is to say, they must not have such irregular faces as to roll, or, in technical terms, be winding to each other. The greatest care must be taken by the workman to have his mortar of a proper consistence,—not too thin, as in drying it would shrink from the work,—nor too thick, for that would prevent the stones from bedding properly. The best way in irregular masonry, or in that composed of small stones thrown as it were between the regular work, as in compound masonry, is to saturate fresh lime with water, and, while hot, to pour it on the work, which hardens and consolidates the whole into one solid mass. This method is much used in joining soft stones, and

brickwork, and is calculated to promote the strength and solidity of the work.

2. Joining by cramps. Cramping is performed, by inserting into the two pieces of stone which are to be bound together, a piece of iron or some other metal, the ends of which, bent at right angles, are inserted in a cavity cut in each stone, the cavities being so large as to admit the iron easily; melted lead is then poured in, to fill the vacant space, and, when cold, a chisel is driven into it, so that it may press close to the work; for all metals expand by fusion, and obstacles may prevent them from contracting in cooling. Cramps composed of copper are, in many cases, very preferable to those made of iron, for they are less likely to oxidize, or rust, or to be affected by the lime or mortar. It would be of advantage to coat the cramps, if made of iron, with some substance that would defend them from the effects of damp. We may here remark, that the channel made to receive the cramp should be dove-tailed, to prevent the lead from coming out, which it is otherwise apt to do, in the course of time. The only objection to the use of copper cramps, in preference to iron, is their expense, which in large public works is not of any importance, and, for common purposes, iron answers very well; but the more malleable or tough the iron, the better it is, as it is more calculated to resist the different temperatures to which the work may be exposed.

3. Joining by joggles. The method of securing the joints of masonry by means of joggles, is chiefly adopted for securing the joints of columns or pillars; and consists in sinking a cavity in the two pieces in such a manner as to make them correspond with each other, and inserting in that cavity a piece of metal, stone, or even wood, so that any lateral thrust may not be able to

separate them. This method may, with very great
advantage, be applied in the construction of domes, and
works of that nature, where it is necessary to avoid the
lateral thrust as much as possible.

We may here take the opportunity offered to us, of
mentioning a plan proposed by Dr. Hutton, in his edition
of Oznamare's Mathematical Recreations,* for taking
away the lateral thrust of domes and cupolas.   The following is the problem proposed, and the solution given.

"How to construct a hemispherical arch, or what the
architects call an arc en cul-de-four, which shall have
no thrust on its piers.

"Let A B, *fig.* 40, be two contiguous voussoirs, which
we will suppose to be three feet in length, and eighteen
inches in breadth. Cut out, on the contiguous sides, two
cavities, in the form of a dove-tail, four inches in depth,
with an aperture of the same extent *a*, *b*, five or six
inches in length, and as much in breadth.   This cavity
will serve to receive a double key of cast-iron, as shown
in *fig.* 41, or of common forged iron, which is still more
secure, as it is not so brittle.   These two voussoirs will
thus be connected together in such a manner that they
cannot be separated without breaking the dove-tail at
the re-entering angle ; but, as each of its dimensions in
this place will be four inches, it will be easily seen that
an immense force would be required to produce that
effect ; for we are taught, by well known experiments
on the strength of iron, that it requires a force of four
thousand five hundred pounds to break a bar of forged
iron, an inch square, by the arm of a lever of six inches ;
consequently, two hundred and eighty-eight thousand
pounds would be necessary to break a bar of sixteen
square inches, like that in question.   Hence there is

* Vol. III. p. 341.

reason to conclude, that these voussoirs will be connected together by a force of two hundred and eighty-
eight thousand pounds; and as they will never experience an effort to disjoin them, nearly so great, as might
easily be proved by calculation, it follows that they
may be considered as one piece.

They might be still further strengthened, in a very
considerable degree, for the height of these dove-tails
might be made double, and a cavity might be cut in the
middle of the bed of the upper voussoir, fit to receive it
entirely: the dove-tail could not then be broken without
breaking the upper voussoir also; but it may be easily
seen that, to produce this effect, an immense force would
be required.

The second method proposed by Dr. Hutton, is more
properly by the aid of joggles. Let A and B, *fig.* 41, be
two contiguous voussoirs, and c *fig.* 42, the inverted
voussoir of the next course, which ought to cover the
joint between A and B. Each of the voussoirs A and B
being divided into two parts, as *a b* and *c d;* then, if at
a b and *c d* we sink a hemispherical cavity, in which to
introduce a globe of very hard marble, and in the upper
voussoir, *fig.* 41, we sink similar cavities, *b c;* this,
when laid on *b c, fig.* 42, will form a secure joint, without any lateral thrust; and the two forces cannot be
separated without a force adequate either to break the
solid stone, or to disunite the marble globe; a force almost
inconceivable, or, at least, one far superior to that produced by the arch; the whole dome, or cupola, is, in
fact, one solid mass, and can exert no lateral thrust
upon the walls on which it is raised. Marble globes
are recommended, because iron is liable to rust; but, if
the joggles were made of iron, and covered with pitch,
before they were placed in the cavities, there would be

little to fear from rust; and particularly as the iron is inclosed in the substance of the stone, and quite excluded from the action of atmospheric causes.

Little need be said in this place as to morticing and tenoning, or dove-tailing, except that they differ slightly from the same operations in joiners' work; for, as cement is used in the joining, they need not be so accurately cut, and are made shorter and thicker than those formed by the joiner, it being sufficient that the parts of each piece to be joined enter into each other five or six inches at most, even in large masses of stone. In small pieces, an inch, or an inch and a half, is sufficient; for, if the tenon or dove-tail be too long, it will decrease the solidity of the joint. For greater security, a small channel is frequently cut in the shoulder of the joint, and melted lead is poured into it, which, filling up the space round the tenon or dove-tail, makes the joint more secure, and the work firm and solid.

In laying some sorts of stones, particularly Portland, Bath, and Gloucester, it is desirable, as far as possible, to place them in the same direction as they had when in the quarry, or, as it is termed by workmen, bedways of the stone; for, if laid in other directions, they are liable to peel and split by the action of the atmosphere.

Having taken a general view of the materials employed by the mason, and the manner in which he uses them, we may proceed to explain the manner in which he polishes his work, or removes those stains and injuries to which highly-polished marbles are subject.

How to clean or polish Marble.

If the stone or marble be rough from the tool or the saw, it will be necessary, first of all, to smooth the surface, by rubbing over it, backwards and forwards, another piece of stone, which is usually fixed to a wooden handle, to give the workman a greater power over it. The polisher begins by sprinkling over the work a coarse grit, moistened with water, as required; this is washed off when a regular though rough surface has been obtained; and the process is repeated with a finer sand, till all the marks of the previous rubbing are removed; a still finer sand, or some fine emery powder, may then be used, and the process continued until a perfectly level surface, fit for polishing, is obtained. The first process in polishing, is to rub the surface with fine flour of emery, by means of a piece of felt, fixed on a board with a weight attached to it. A thick cloth, by workmen called fearnought, is sometimes substituted for the felt, and is much better. After this operation, take a fresh piece of felt, and apply putty-powder in the same manner as the emery had been before used; first with water, if it should be necessary, and then without, which will produce a fine polish. Some workmen use rotten-stone, others tripoli, and finish with fine flour, on a piece of buff leather.

Another method of polishing is that called the Italian. After the work has been levelled with sand and water, as already described, it is finished with a piece of lead, having a surface that corresponds with that of the work to be polished. Beginning with the coarser emery, the workman proceeds, by degrees, to those which are finer; and finishes with calcined tin, and a piece of leather.

When the marble is very hard, and is capable of bearing a very high polish, another method may be adopted. After taking out all the marks left by the stone and sand, the workman may use a fine pumice-stone, and rub it until every scratch disappears, and then polish it in the usual way with tripoli powder. After this has been done, to give it a higher gloss, prepare a tool of the shape required, from a piece of lime-tree wood, and on it spread, evenly, a coat of pitch, moistened with a few drops of vinegar, and a powder made in the following manner: four parts of tripoli, with one of blue-vitriol, both ground very fine. When the polish is nearly obtained, fresh powder must not be added to the tool. In this manner, if properly managed, a polish equal to that of a mirror may be obtained, and, although more troublesome than other methods, yet the effect produced amply repays for the care and trouble bestowed.

To Clean Marble.

All marbles; and especially the statuary and light veined ones, are very liable to be stained, having a natural tendency to imbibe the colouring matter of vegetable and mineral substances. Even when a polished marble is packed in hay or shavings, it is by no means safe, but is in much danger of being spoiled. For want of sufficient care, chimney-pieces and other ornamental marbles are very frequently stained and discoloured; it may, therefore, be desirable that we should explain some of the methods by which these injuries may be removed.

To clean marble, mix quick-lime with soap-lees, so as to form a mixture having the consistence of cream, and apply it immediately with a brush. If this composition be allowed to remain for a day or two, and be then washed off with soap and water, the marble will appear as though it were new.

To extract grease or oil from stone or marble, make a strong lye of pearl-ash and water, and adding unslacked lime, allow it to settle, and pour it off for use; or it may be kept for a long time if placed in a bottle and well corked. Place a little upon any grease spot, and after it has remained for a few minutes, wash it off with clean water.

Stains may frequently be taken out by a very simple process, but it does not always succeed, and then one or the other of the former methods may be tried. Take any quantity of whitening and mix it with good soap-lees, until it has the consistency of cream or thin paste; then lay it evenly on the stained part with a brush, and, after it has remained for a few days, wash it off, and repeat the process if the stain be not quite removed.

Iron-mould and ink-spots may be taken out in the following manner: take half an ounce of butter of antimony, and one ounce of oxalic acid, and dissolve them in a pint of rain water; add flour, and bring the composition to a proper consistence. It may be applied in the same manner as the composition already described.

Cements.

We have already spoken of cements used by the brick-layer, and as the same are employed by the mason, we refer the reader to that part of the work in which they are described. There are, however, some few which are serviceable to the mason, and are not employed by the bricklayer; to these it will be necessary to refer in this place.

A delicate cement for small work may be made in the following manner: take half a pint of milk, and when it is near boiling, add vinegar, until a curd is formed; then strain off the whey, and add it to the white of four or five eggs, and when well mixed, sift quick-lime into

L

it, stirring it all the while, until it has the consistency of paste. For small work, and for joining pieces broken off, this cement is well suited, for it resists the action of both fire and water.

There is a hot cement which is very useful for stopping flaws or holes, and may be made of the same colour as the marble or stone, by mixing a colour with it, or the powder of the stone itself. It is also used for veneering, or fixing costly marbles on those less valuable, and for inlaying and mosaic work. It may be made in the following manner: melt half a pound of bees'-wax and a quarter of a pound of powdered rosin together in a pipkin, and to these add an ounce of finely powdered chalk, and an ounce of fine brick-dust or sand, of the colour required. Let the whole composition simmer together for a quarter of an hour, keeping them constantly stirred, and use the composition while hot. The stones to be cemented, must be moderately warmed before the cement is applied.

To make a suitable cement for small work, take any quantity of oyster shells and calcine them, and grinding them very fine, sift the powder through a piece of fine muslin; to this add the whites of a sufficient number of eggs to form a paste.

THE PLASTERER.

In speaking of the work executed by the plasterer, we must refer to some of the most important facts which relate to the spreading, evenly and smoothly, on the external and internal surface of walls, and on ceilings, a composition known as mortar or cement. The plasterer has also to execute all those decorative and ornamental parts of a building, which are intended to imitate statuary, or carving; and these require, not only the use of well selected materials, but also an accurate execution. If there be one department of the art of building in which the modern style exceeds the ancient, it is in plasterer's work, for it has risen to that excellence which may almost warrant us in calling it perfect. Interior work is so executed, that, with the aid of colour, it resembles stone; and the stucco which is applied to the exterior, if properly executed, rivals the more solid material in beauty as well as in durability.

The plasterer's work differs very much from the works already described, for the workman has no occasion to study the influence of his work upon others, so far at least as strength is concerned. His duty is to cover the naked timbers and brickwork in ceilings and walls, and to give such a face to his work as shall be suited to the painter or the paper-hanger. Supposing him to be a good workman, there are only two things which require his attention; the purity of his materials, and the accuracy of his tools. The plasterer must be careful that all his trowels, and stopping and picking-out tools, be cleaned after use, so that rust does not form upon their face and injure the work; and that his straight-edges and moulds be fit to execute the work for which they were intended.

We may here mention, that it is of some importance to the plasterer to be acquainted with the art of designing ornaments belonging to his own work, as mouldings, foliage, and figures; and it is even more important for him to acquaint himself with the art of modelling in clay, which is one of the greatest assistants he can engage. The art is, in itself, worthy of attention, and would be highly valuable to a person who only sought to obtain, by its means, amusement and a knowledge of forms; but the plasterer will find it useful, for it will give him a readiness in finishing the returns and mitres of his work, where moulds cannot be used. All that we can do in this chapter is to explain the several compositions, and the manner in which they are used, for the purpose of assisting the young workman as much as possible.

Coarse Stuff.

Coarse stuff, or lime and hair, as it is sometimes called, is prepared in the same way as common mortar, with the addition of hair procured from the tanner, which must be well mixed with the mortar by means of a three-pronged rake, until the hair is equally distributed throughout the composition. The mortar should be first formed, and when the lime and sand have been thoroughly mixed, the hair should be added by degrees, and the whole so thoroughly united, that the hair shall appear to be equally distributed throughout.

Fine Stuff.

This is made by slaking lime, with a small portion of water, after which so much water is added as to give it the consistence of cream. It is then allowed to settle for some time, and the superfluous water is poured off, and the sediment is suffered to remain till evaporation reduces it to a proper thickness for use. For some kinds of work it is necessary to add a small portion of hair.

Stucco for inside of Walls.

This stucco consists of the fine stuff already described, and a portion of fine washed sand, in the proportion of one of sand to three of fine stuff. Those parts of interior walls which are intended to be painted are finished with this stucco. In using this material, great care must be taken that the surface be perfectly level, and to secure this it must be well worked with a floating tool or wooden trowel. This is done by sprinkling a little water occasionally on the stucco, and rubbing it in a circular direction with the float, till the surface has attained a high gloss. The durability of the work very much depends upon the care with which this process is done, for if it be not thoroughly worked, it is apt to crack.

Gauge Stuff.

This is chiefly used for mouldings and cornices which are run or formed with a wooden mould. It consists of about one-fifth of plaster of Paris, mixed gradually with

L *

four-fifths of fine stuff. When the work is required to set very expeditiously, the proportion of plaster of Paris is increased. It is often necessary that the plaster to be used should have the property of setting immediately it is laid on; in all such cases gauge-stuff is used, and consequently it is extensively employed for cementing ornaments to walls or ceilings, as well as for casting the ornaments themselves.

Bailey's Compo.

The plaster or stucco, known under this name, is composed of three parts of Dorking-lime, and one part of fine, washed, river sand. These ingredients are well mixed together in a dry state, and put into casks, to prevent the access of the air. When required for use, it is first mixed with water, to the consistence of thick whitewash, and applied with a stiff brush as a ground, preparatory to spreading the wall with a mortar of sufficient thickness. The mortar is floated, that is, well rubbed with the wooden float, or the trowel, sprinkling it occasionally with water, till the surface is quite smooth and level.

Higgins' Patent Stucco.

The stucco, invented by Dr. Higgins, is seldom if ever employed, as much from the trouble as from the expense of making it. To fifteen pounds of the best stone lime add fourteen pounds of bone-ashes, finely powdered, and about ninety-five pounds of clean, washed sand, quite dry, either coarse or fine, according to the nature of the work in hand. These ingredients must be intimately mixed, and kept from the air till wanted. When required for use, it must be mixed up into a proper consistence for working, with lime-water, and used as speedily as possible.

Parker's Cement.

This cement, which is perhaps better than any other for stucco, as it is not subject to crack or flake off, is now very commonly used, and is formed by burning argillaceous clay in the same manner that lime is made; it is then reduced to powder, by the process described in a previous part of this work. The cement, as used by the plasterer, is sometimes employed alone, and sometimes it is mixed with sharp sand; and it has then the appearance, and almost the strength, of stone. As it is impervious to water, it is very proper for lining tanks and cisterns.

Hamelin's Cement.

This cement consists of earthy and other substances insoluble in water, or nearly so; and they may be either those which are in their natural state, or which have been manufactured, such as earthenware and china; those being always preferred which are least soluble in water and have the least colour. When these are pulverized, some oxide of lead is added, such as litharge, grey oxide, or minium, reduced to a fine powder; and to the compound is added a quantity of pulverized glass or flint stones; the whole being thoroughly mixed and made into a proper consistence with some vegetable oil, as that of linseed. This makes a durable stucco or plaster, which is impervious to wet, and has the appearance of stone.

The proportions of the several ingredients are as follows: to every five hundred and sixty pounds of earth, or earths, such as pit sand, river sand, rock sand, pulverized earthenware, or porcelain, add forty pounds of litharge, two pounds of pulverized glass or flint, one pound of minium. and two pounds of grey oxide of lead. Mix the whole together, and sift it through sieves of different degrees of fineness, according to the purposes to which the cement is to be applied.

The following is the method of using it. To every thirty pounds weight of the cement in powder, add about one quart of oil, either linseed, walnut, or some other vegetable oil, and mix it in the same manner as any other mortar, pressing it gently together, either by treading on it, or with the trowel; it has then the appearance of moistened sand. Care must also be taken that no more is mixed at one time than is required for use, as it soon hardens into a solid mass. Before the cement is applied, the face of the wall to be plastered should be brushed over with oil, particularly if it be applied to brick, or any other substance that quickly imbibes the oil; if to wood, lead, or any substance of a similar nature, less oil may be used.

Maltha, or Greek Mastick.

This is made by mixing lime and sand in the manner of mortar, and making it into a proper consistence with milk or size, instead of water.

Wych's Stucco.

Take four or five bushels of such plaster as is commonly burnt for floors about Nottingham, or the same quantity of tarras, plaster, or calcined gypsum, and beat it into a fine powder. Then sift it into a trough, and mix with it one bushel of pure coal ashes, well calcined; pour water upon it gradually, until the whole mass has the consistence of mortar.

Plaster in imitation of Marble.

The plastering to which we refer is called scagliuola, and was introduced into this country, from France, by the late Mr. Holland. This species of work is exquisitely beautiful when done with taste and judgment, and is so like marble to the touch, as well as in appearance, that it is scarcely possible to distinguish the one from the other. We shall endeavour to explain its composition, and the manner in which it is applied; but so much depends upon the workman's execution, that it is impossible for any one to succeed in an attempt to work with it, without some practical experience.

Procure some of the purest gypsum, and calcine it until the large masses have lost the brilliant sparkling appearance by which they are characterized, and the whole mass appears uniformly opaque. This calcined gypsum is reduced to powder, passed through a very fine sieve, and mixed up, as it is wanted for use, with Flanders glue, isinglass, or some other material of the same kind. This solution is coloured with the tint required for the scagliuola; but when a marble of various colours is to be imitated, the several coloured compositions required by the artist must be placed in separate vessels, and they are then mingled together in nearly the same manner that the painter mixes his colour on the pallet. Having the wall or column prepared with rough plaster, it is covered with the composition, and the colours intended to imitate the marble, of whatever kind it may be, are applied when the floating is going on.

It now only remains to polish the work, which, as soon as the composition is hard enough, is done by rubbing it with pumice stone, the work being kept wet with water, applied by a sponge. It is then polished with Tripoli and charcoal, with a piece of fine linen, and finished with a piece of felt, dipped in a mixture of oil and Tripoli, afterwards with pure oil.

Composition.

This is frequently used instead of plaster of Paris, for the ornamental parts of buildings, as it is more durable, and becomes in time as hard as stone itself. It is of great use in the execution of the decorative parts of architecture, and also in the finishings of picture frames, being a cheaper method than carving, by nearly eighty per cent.

It is made as follows: two pounds of the best whitening, one pound of glue, and half a pound of linseed oil, are heated together, the composition being continually stirred until the different substances are thoroughly incorporated. Let the compound cool, and then lay it on a stone covered with powdered whitening, and heat it well until it becomes of a tough and firm consistence. It may then be put by for use, covered with wet cloths to keep it fresh. When wanted for use it must be cut into pieces, adapted to the size of the mould, into which it is forced by a screw press. The ornament, or cornice, is fixed to the frame or wall with glue, or with white lead.

Lime-Wash.

As this is the most common wash for walls, and is at the same time the cheapest, it is generally used for common work. But a very superior whitewash may be made with it, if proper care be taken in its preparation, for it is less apt to peel than those which are made with whitening and size. The following process will be found to answer the purpose: Take a sufficient quantity of the best lime, in small pieces, and pour clean water upon them, stirring the liquid for some time. Then let the solution remain for a few minutes, and pour it off into another vessel, leaving the heavy particles behind. Add more water, stirring it as before, and leave it again to settle. Then pour off the water from the top, strain the whole through a very fine sieve, and keep it covered until wanted for use, when a sufficient quantity of water to reduce it to the proper consistence may be added. In using lime-wash, it is better to put two thin coats on a wall than one thick one, for the first coat has often a smeary and uneven appearance. With these precautions, a very superior lime-wash may be made, fit to be used for any kind of work, and not liable to the faults of the common wash. It is, however, necessary that care should be taken as to the cleanness of the wall or ceiling to which it is applied, and especially that it be not applied over a coat of size, for then it is almost sure to turn yellow.

PLASTERING.

There are two ways in which internal plastering may be executed,—on laths, or on walls; in the latter case the laying on of the first coat is called rendering. With one or two remarks upon the manner of executing the work, according to circumstances, we shall close this chapter.

It has been already stated, that plastering is sometimes formed upon laths. Laths differ in size, and in the quality of the wood of which they are formed. Laths are, in building, distinguished by their thickness; there is the single lath, the lath and a half, and the double lath, the single lath being about a quarter of an inch thick. The most serviceable lengths are three and four feet. The single lath is commonly used for partitions, the double lath for ceilings. Laths, of all kinds, are made of both Baltic and American fir, and of oak, the former being most commonly employed.

The process of spreading the first coat of lime and hair over the partition or ceiling to be plastered, is called laying. Floating is that process by which a surface of plaster is made perfectly plane, by means of an instrument called a float. Setting is a finishing process; in the best work gauged stuff is used, and in common work fine stuff. Sometimes three coats of plaster are placed on a wall or ceiling, and sometimes only *two*, so that the setting coat may be either the second or third.

Lath, laid and set, is a phrase, in plastering, signifying two-coat work; the coat of hair and lime with which the laths are covered, and the coat of plaster mixed with a little plaster of Paris, which is floated for the purpose of obtaining a smooth surface. When the first coat is laid on, it is sometimes worked over with a lath so as to form a key for the next coat; this is called pricking up.

Lath, prick up, float and set for paper, is three-coat work; pricking up, floating, and finishing for paper.

Rendering, is the first coat upon a naked wall; thus we say, rendered and set; that is, a coat of coarse stuff on the naked wall, and a coat of fine stuff upon it.

Render, float, and set, is three-coat work; the first process is laying on a coat of lime and hair; the second floating a coat of the same composition, except that a little more hair is added; and the last a coat of fine stuff with white hair.

These general remarks upon the composition of the several cements used by the plasterer, and the terms by which they are known, according to circumstance, will, it is hoped, assist the student. But he will perhaps forgive a remark by which he may profit, if he should be induced to take our advice. It is an unworthy pride, to prefer ignorance rather than expose it by asking for information. We have known those who have been placed in circumstances most advantageous for the accumulation of knowledge, but from a false feeling of delicacy, or from the indulgence of that haughty and proud spirit so often perceived in young persons, though so unworthy their years, and alike disgusting to their superiors and inferiors, who have never availed themselves of their advantages, but have been as ignorant of the meaning of terms at the end of five years' apprenticeship, as they were, at the commencement, with the terms themselves. We chiefly refer, in these remarks, to the architectural student,—and what is the result of all? The time arrives when the student becomes a teacher, and is required to superintend the engagements

M

of all those who are employed in the construction of buildings; and he finds himself absolutely incapable of the task. He has perhaps made a beautiful set of drawings, and has arranged for the construction of every part, in the same way as his predecessor was accustomed to do; but he can neither tell why he has done it in this way, in preference to any other, nor discover any deviation that may be made from his orders. We have not drawn a fanciful picture, and we warn the student, lest he should fall into the same error, and suffer the same inconvenience as many have done before him.

THE PLUMBER.

THE plumber is chiefly engaged in the execution of such works, in the art of building, as require to be formed of lead: but, within the last four years, zinc has been extensively employed instead of lead, and the plumber has undertaken the execution of such works. This substance, however, is not so durable or malleable as lead, but it is liable to crack, and especially if it be fastened. It is not, therefore, so desirable a material for building purposes, but it is cheaper, and may be advantageously employed for many purposes. We shall, however, in this place, chiefly direct our attention to a consideration of the uses and properties of lead, a substance that must always be extensively employed in building.

Lead has a blueish white colour, and its face when first formed has a bright, glittering appearance, which is soon tarnished by exposure to the air; and losing its lustre, it acquires a dull greyish colour, an effect resulting from the oxidation of the metal; that is to say, the metal combines with a proportion of the oxygen of the air, and the coat which covers the surface of the metal is an oxide. Every one who has examined a lead flat after it has been laid a few months, or the interior of a cistern, or water-pipe, must have observed this effect, and it is worthy of remark that, although water does not oxidize lead, yet it accelerates the effect of atmospheric air.

Lead is found native in the state of a sulphuret, that is to say, combined with a certain proportion of sulphur. In this state it would not be suited for the purposes to which it is now applied; and, consequently, the manufacturer is compelled to free it from the ingredient with which it is combined, and from the earthy minerals and other impurities. This is done by the process of roasting. The ore is first of all broken into pieces and washed, and then placed in a reverberatory furnace, where it is exposed to an intense heat. The sulphur is in this manner sublimed, and the metal itself carried off into moulds. Each mould contains one hundred and fifty-four pounds, and is called a pig of lead.

But the plumber uses lead chiefly in sheets, which, in many cases, are made by himself from the pig-lead, and from the old material which he purchases or takes in exchange. There are, however, two kinds of sheet-lead, cast and milled; and it is the cast which is made by the plumber. All sheet-lead is valued according to its thickness, that is to say, according to the number of pounds contained in every square foot; and architects, when they describe the kind of lead to be employed, say five, six, or seven pound lead, according to the thickness required for the particular kind of work. In the preparation of specifications, milled lead is generally provided for, but the plumber often lays cast lead instead; but this is not the only manner in which the plumber deceives the architect, for he often puts a lead of less weight than is contracted for, and, as it requires considerable practice in order to detect, by the feel, whether the lead is as heavy as was required, he practises the deception almost without a chance of detection. The milled lead is not made by the plumber, but is purchased of the lead manufacturer, for it requires a particular apparatus for its preparation.

Sheet-lead is chiefly used to cover the flats of roofs, gutters, and cisterns; and for flashings.

Leaden pipes, used to carry water from roofs, and for water-works generally, were at one time almost constantly made of sheet-lead bent round a wooden staff, of the size required for the bore; and the joints were united by solder. This plan, however, was not found to succeed so well as was desired, and they are now cast upon a cylindrical iron mould. Pipes are described by their bore, thus we speak of one, two, or three inch, lead pipe; but nearly all plumber's work is estimated by the weight.

Lead should be laid with as few joinings as possible, but it is quite impossible to avoid them altogether;—and there are two methods in which they may be executed: First by lap or roll joint, which should always be preferred, and secondly by solder. Solder is a metallic alloy, used to unite together the edges of some metallic substance; and there is one principle that should always govern its composition,—it must melt more readily than the substances to be joined. The solder employed by the plumber is made of equal parts of tin and lead, and is run into the joint in a liquified state; after which it is smoothed down by a grozing-iron heated almost to redness, and finished off by filing or scraping. It has been already stated that lead is particularly subject to oxidation, and, to prevent this in the process of soldering, the edges of the joint are scraped clean and covered with borax, which defends the lead when the heat is applied. If zinc be employed instead of lead, all the joints must be formed by laps, and not by soldering, so as to give it a freedom of expansion and contraction.

M *

THE PAINTER.

THE painter covers with oil-colour much of the joiner's, plasterer's, and smith's work. The art of house-painting is very ancient; and, when first introduced, its only object was, in all probability, decoration; but it has now another object, the preservation of materials. There are few woods that will long remain sound, if exposed to a constant change of weather; alternate wet and dry soon causes a piece of timber, however sound, to crack; and encourages the dry-rot. Now, in the erection of buildings, there must always be some materials thus exposed, and to prevent the effect, they are covered with a coat of paint. Iron-work also, if exposed to the varia- tions of weather, will decay by oxidation; that is to say, it will rust; but, when covered with oil-paint, it is pre- served from this effect, and will last an indefinite period. It will therefore appear that, if we merely considered the durability of buildings, the painter would be an im- portant person, and particularly so in so variable a climate as that in which we live.

But there are other situations in which the materials are not thus exposed to decay, and paint is applied to improve the appearance of the work. We are not among those who prefer an imitation to a reality, and would cover a fine-grained wood, as many do. No painting can equal the old wainscot we sometimes find in the mansions of our ancestors; but the fine-grained woods are now very expensive, and, if they could be obtained,

would not be suitable for all situations. It is therefore
necessary, that an inferior material, so far at least as
its appearance is concerned, should be employed, and
that such colours should be given to it as may be adapted
to the purposes of the apartment in which they are used.
The present excellence in this art has done much to
improve the appearance of our domestic architecture,
and to provide those elegancies, at a moderate price,
which were once obtained by the wealthy only. This
department of building, therefore, demands our careful
attention, and we shall endeavour to put the reader in
possession of some practical information concerning it,
for it is only this that is within our reach. We cannot
teach the harmony of colours, or their appropriateness to
particular situations; but there are principles to be fol-.
lowed, although it is often said that the choice of colours
is a matter of indifference.

MATERIALS.

Paints are made of various mineral productions, which
are ground, that is, reduced to powder, and then made.
liquid by some fluid, so as to admit of application with a
brush. The colouring substance is sometimes ground
in water, and then a size must be added, to give it a
stronger adhesive power: sometimes it is mixed with
spirits of wine, and as this fluid evaporates readily, only
a small quantity must be mixed at a time: but it is.
commonly ground in oil, and mixed with turpentine,
or turps, as it is called by workmen, a substance obtained from larch and fir trees. Ceilings, and the
stringing of staircases, are frequently painted in water-
colours; but woodwork, and the walls of rooms, are
commonly worked in oil-colours. In that kind of

finishing called flatting, because it makes a very even
and dull surface, the colour is prepared-with turpentine
only ; and as the execution of the work is very readily
detected when it is flatted, great care should be taken
in the process, and a clever expert workman should be
employed.

A Preparation for painting Ceilings.

Take a sufficient quantity of Spanish white, and,
having pounded it, let it soak in water for about two
hours. To give it a more or less dark tint, as may be
required, charcoal should be infused in water, and added
to the composition; and, to give it the adhesive property, strong size,—a larger quantity being required
when used on new work, than when the same preparation has been applied at some former time. If a ceiling
has been before whitened, it is generally necessary to
scrape off the former coat before the commencement of
the work.

To whiten internal Walls.

A very superior material for the whitening of internal
walls may be made in the following manner. Take a
quantity of very fine lime, and, passing it through the
finest sieve that can be obtained, place it in a vessel
sufficiently large for the purpose; then filling it with
water, thoroughly mix the lime and water with a wooden
instrument, so as to diffuse the whole of the solid material through the fluid. When this has been done, let the
mixture stand for about four and twenty hours, so that
the lime may be deposited, and then draw off the liquid,
which will contain the impurities previously mixed with
the lime. Fill the vessel again with water, and mix the

ingredients as before, drawing off the water when the sediment has been formed. The lime will then remain at the bottom of the vessel, and the impurities being withdrawn, it will be exceedingly white; so bright indeed, that it will be necessary to add a little Prussian blue. When the purified lime is mixed with turpentine, size, and a very small quantity of alum, a composition will be formed, which, when applied to the face of the work, will have a peculiarly beautiful appearance. The work will be greatly improved by rubbing it with a brush, not so stiff as to scratch it, but sufficiently so as to produce a strong friction.

To paint on Stucco.

Great care is required in painting upon stucco, for the work must be not only thoroughly dry, but free from any liability to dampness; that is to say, the walls themselves must be dry. It is, consequently, usual to allow the stucco to remain for several months before it is painted; and this is especially necessary when it covers over a large surface, as in the walls of churches, chapels, and theatres. If the paint be applied too soon, the work will have a blotched appearance, and be probably filled with small vesicles, formed during the evaporation of the water. When the work is dry, it may be prepared by covering it with a coat of linseed oil, boiled with dryers. This must be laid on very carefully, or the face will be irregular. The colour may then be applied, and four coats will not be too much, the work being new. Persons are generally so anxious to have their buildings finished, that they disregard the future appearance of the work; and, within a few weeks after the application of the stucco, cover it with paint. But it would, in all cases, be sufficient to wash the surface with distemper, as it would give a finished appearance to the building, and make it less necessary to hurry the work. But when it is sufficiently dry to receive the oil-colour, the water-colour, that is to say, the distemper, should be removed, which may be done by washing; and as the water does not penetrate into the substance of the stucco, it will dry in a few days, and receive the oil-colour. The tints may be regulated by mingling different colours, as in all other kinds of painting.

GRAINING.

The art of imitating the grain of the more expensive woods is now brought to so great a degree of perfection, that it is often almost impossible to determine, without feeling the surface, whether we are looking upon the wood, or an imitation of it. Mahogany, satin-wood, rose-wood, maple, and some others, are frequently imitated; and it is but seldom that a good house is finished without the introduction of some graining. Doors to drawing-rooms, dining-halls, and passages, are usually painted, if some handsome grained wood be not introduced. The dado, and skirtings, are also frequently finished in this manner. But it is not now so commonly employed as it was a few years ago. Delicate party-colours are often preferred for drawing rooms, and for those apartments which are most frequently inhabited. The process of graining is very simple. The workman first prepares the surface with two or three coats of oil-paint, and then forms the ground of the graining, the colour of the ground being regulated by the colour of the wood to be imitated. If, for instance, it be required to imitate satin-wood, then the ground will be formed of Naples yellow and ceruse, worked up with turpentine. When this coat is perfectly dry, the graining is commenced, the painter preparing small quantities of such colours as he requires, upon his pallet, and applying them with camel's-hair pencils, of different sizes, and flat hog's-hair brushes. When the work is finished, it must be allowed to remain until perfectly dry, and then covered evenly with one or two coats of good oil varnish. The same process is adopted in the imitation of marbles, for chimney-pieces, pilasters, and other ornamental work.

In some cases, graining in distemper may be adopted with great success, although we are not aware that it is much practised. Some time since, having a large surface of woodwork to grain oak, within a period so short as to prevent its execution in oil, as it could not have dried in time, if the graining itself could have been executed by a word of command, we gave an order to finish it in distemper. A clever painter undertook the work; completed it to our satisfaction; and a coat of varnish was then applied. The work has been now completed nearly four years, and it could not, at the present moment, be distinguished from work finished in oil colours. We would, therefore, strongly recommend the process, in all those cases where despatch is necessary, for interior work.

ON COLOURS.

The following remarks on colours are chiefly extracted from De Morveau's paper on that subject :—

" White is the most important colour in painting, for it is to the artist the material of light, which he is

required so to distribute over his work as to bring his
objects together, and to give them relief; and this it is
which is the magic of his art. For these reasons, I
shall at present confine my attention to this colour.

" The first white that was discovered, and indeed the
only one yet known, was extracted from the calx of lead.
The danger of the process, and the dreadful distemper
with which those employed in it are often seized, have
not yet led to the discovery of any substance that can be
used in its place. There has, indeed, been less anxiety
about the artist, than the perfection of the art ; and the
manufacturer, guided by this, has varied the preparation,
to render the colour less liable to change; and hence
the different kinds of white,—the white of crems, white
lead in shells, and white ceruse. But every person,
conversant in colours, knows that the foundation of all
these is the calx of lead, more or less pure, or more or
less loaded with gas. That they all participate of this
metallic substance will indeed appear evident from the
following experiment, which determines and demon-
strates the alterability of colours by the phlogistic
vapour.

" I poured into a large glass bottle, a quantity of liver
of sulphur, on a basis of alkali, fixed or volatile it makes
no difference; I added some drops of distilled vinegar,
and I covered the mouth of the bottle with a piece of
pasteboard cut to its size, on which I disposed different
samples of crems, of white lead, and of ceruse, either in
oil or in water. I then placed another ring of pasteboard over the first, and tied above all a piece of bladder,
round the neck of the bottle, with a strong packthread.
It is evident that, in this operation, I took advantage of
the means which chemistry offers, to produce a great
quantity of phlogistic vapour, to accomplish instantane-

ously the effects of many years; and, in a word, to apply
to the colours the very same vapours, to which the
picture or work is necessarily exposed, only more ac-
cumulated and more concentrated. I say the same
vapour, for it is now fully established that the smoke of
candles, animal exhalations of all kinds, alkalescent
odours, the electric effluvia, and light, furnish continually a quantity more or less of matter, not only analogous, but identically the same with the vapour of vitriolic
acid mixed with sulphur.

"If it happens that the samples of colours are sensibly
altered by the phlogistic vapour, then we may conclude
with certainty, that the materials of which the colours
are composed, bear a great affinity to that vapour; and,
since it is not possible to preserve them entirely from it
in any situation, that they will be more or less affected
by it according to the time it stands, and other circumstances.

"After some minutes' continuance in this vapour, I
examined the samples of colours submitted to its influence, and found them wholly altered. The ceruse and
the white lead, both in water, and in oil, were changed
into black; and the white of crems into a brownish
black; and hence those colours are bad and ought to be
abandoned. They may, indeed, be defended in some
measure by varnish, but this only retards for a time the
contact of the phlogistic vapour; for as the varnish loses
its humidity it opens an infinite number of passages to
this subtile fluid.

"There are three conditions," says De Morveau,
"essential to a good colour in painting.

"First. That it dilute easily, and take a body, with
oils and with mucilages, or at least with one or other of
these substances; a circumstance which depends upon

a certain degree of affinity. Where this affinity is too strong, a dissolution ensues; the colour is extinguished in the new composition, and the mass becomes more or less transparent; or else the sudden reaction absorbs the fluid, and leaves only a dry substance, which can never be again softened. But if the affinity be too weak, the particles of colour are scarcely suspended in the fluid, and appear on the surface that is coloured, like sand, which nothing can fix or unite.

" The second condition is, that the materials of which colours are composed, do not bear too strong an affinity for the phlogistic vapour. The experiments in which I submitted whites from lead to this vapour, afford a certain means of ascertaining the quality of colours in this respect, without waiting for the slow impression of time.

"A third condition equally essential is, that the colouring body be not volatile, that it be not connected with a substance of a weak texture, susceptible of a spontaneous degeneracy. This consideration excludes the greater part of substances which have received their tints from vegetable organization; at least it makes it impossible to incorporate their finer parts with a combination more solid."

. These remarks are exceedingly judicious, and show the conditions by which the formation of colours are bounded.

It is very easy to produce the varieties of shade which may be required in house-painting, for the mineral substances employed for this purpose are quite adequate to produce the effect. There is no difficulty in selecting suitable colouring material, but it is not easy to obtain a good and durable white. The objections to the use of lead have been already stated, and we would now recom-

mend the following observations to the careful consideration of manufacturers and workmen. There are two
objections against the use of lead; it is unsuited to the
purpose to which it is applied, so far at least as its in-
stability of colour is considered; and it is detrimental to
the health of all those who are employed in its manufacture and in using it; we might also add to those who
live in rooms in which it has been recently employed.
It therefore becomes a question of some importance,
whether any other substance can be employed that is of
a less injurious quality, and that is equally or more
adapted to the purpose required. In answer to these
questions we quote the experiments of De Morveau, and
his remarks upon them. " I placed in my apparatus
pieces of cloth, on which were laid the white of calcare-
ous tartar in water, and different preparations of white,
from tin and zinc, both in oil and water, and I allowed
them to continue exposed to the phlogistic vapour during
a sitting of the academy; if they were not altered, their
superiority over the whites in use would be sufficiently
established. The sitting continued for near an hour;
and the bottle having been opened, all the colours continued to have the same shade as they had before. I
can therefore recommend to painters those three whites,
and particularly that of zinc, the preparation of which is
exposed to less variation, the shade more lively and
uniform, and moreover it is fit for all purposes, and
perhaps procured at less expense.

" I will assert farther, that it may be procured in sufficient quantities to supply the place of ceruse in every
branch of the art, even in interior house-painting. I
would recommend it, less with the view of adding new
splendour to this kind of ornament, than for the safety of
those who are employed in preparing or using it, and

perhaps for the safety of those who inhabit houses ornamented in this manner.

"But without being too sanguine, although the process in the fabrication be simplified in proportion to the
demand, as is usually the case, yet there is reason to
apprehend that the low price of ceruse will always give
it the preference in house-painting.

"M. Courtois, connected with the laboratory of the
academy, has already declared that it is used for house-
painting, less, however, in regard to its unalterability,
than to its solubility; and this can be the more readily
accounted for, as the flower of zinc enters into many of
the compounds of the apothecary. The same gentleman
has arrived at the art of giving more body to this white,
which the painters seemed to desire, and also of making
it bear a comparison with white lead, either in water or
oil. The only fault found with it, is its drying slowly
when used in oil; but some experiments which I have
made incline me to believe that this fault may be easily
remedied, or at least greatly corrected, by giving it more
body. At any rate, it may be rendered siccative at pleasure, by adding a little vitriol of zinc, or copperas slightly
calcined.

Painters already know the properties of this salt, but
perhaps they do not know that it mixes with the white
of zinc better than with any other colour; the reason is
they have chemically the same base. It is prepared by
purging the white copperas of that small portion of iron
which would render it yellow, and this is easily done in
digesting its solution, even when cold, on the filings of
zinc. The mixture of this salt, thus prepared, is made
on the palette, without producing any alteration, and a
small quantity will produce a great effect.

General Remarks.

It is commonly said that any body can execute all that is required of a house-painter. This assertion, however, cannot be substantiated; it is not so easy to prepare and apply a coat of paint, in a "workman-like manner," as some may imagine: it is still less easy to paint in party-colours; and very few can produce a good piece of graining. But the painter should not only be acquainted with the method of applying the paint, when it is provided for him, and the brush placed in his hand, but he should know the composition of the colours; the manner in which they are made; and the colours which most harmonize with each other when they are associated together.

N *

THE SMITH.

Iron is a very important material in the art of building, on account of its great tenacity and capability of resisting strains. In the present day it is very extensively employed; perhaps modern architects have gone to an excess in this matter, for in the practice of some, iron is so much a *sine qua non*, that it solves all difficulties and covers excessive ignorance. If a professional man should find some difficulty in designing a truss suited to carry a determined or undetermined weight, it is very easy to introduce an iron one. We do not say that such a practice is common in the profession, but we believe that it more frequently directs the use of this material than is commonly admitted. There are many buildings, the framework of which is, in fact, composed of iron; but we cannot consider this judicious. In some places the metal may be introduced with great advantage, but there are others in which other materials would serve not only as well but very much better. There is one thing to be considered in the use of iron, which seems to be frequently forgotten, though it is by no means unimportant;—it has the property of great expansibility by heat, and should not, therefore, be employed in great lengths. We could give many instances in which bressummers, of considerable bearing, have been made with this metal; and those who designed them would justify the plan by telling us, that although iron expands considerably by heat, yet the change of temperature to which it is subject, when used in construction, is so unimportant that it may be altogether neglected in its use. But on the other hand it should be remembered that, when used in considerable lengths, there is an expansion which, however small, acting upon walls, must tend to weaken them. The best illustration that we can give of the great dilatation to which iron is subject at high temperatures, is the application of the principle by a celebrated French architect. The walls of a public building in Paris had spread, or in other words, were thrown out of their perpendicular, and there was some fear of the future safety of the building, so that it became an object to raise them again to their proper position. To effect this, iron bars were carried through the building, from wall to wall, one end being attached to one wall, and the other passing through the opposite, the end being furnished with a screw, and a nut moving upon it. When the nut had been screwed as closely as possible to the wall, a series of lamps were placed under the bar, so as to raise its temperature to a red heat. This caused the iron to expand, and the nut was then screwed up again to the wall; but when it cooled it contracted, and drew up the walls with it. The same process was repeated until the walls were brought to their perpendicular position.

Britain is very advantageously situated for procuring iron; there is, perhaps, no other country in the world that possesses the same resources in so small a space. This is a fact of great commercial importance, for it may be also mentioned, that the metal is found in association with coal, so that in those places where the

material is obtained, there is also the means of smelting
it at a very small expense, and this will account for the
low price at which it is brought into the market.

The quality of iron varies greatly; there are some
kinds which are much less capable of resisting a pressure than others, and are liable to crack when bearing
great weights. There is also a considerable difference
between cast and wrought iron; the latter will bend
before it breaks, is flexible, and yet tough. Cast iron is
very readily broken, will not resist a heavy blow, and
will break rather than bend. It is therefore very desirable to employ wrought iron in buildings, when the metal
is at all introduced, but, on account of it being more
expensive than the cast iron, it is seldom used for heavy
work. Columns, bressummers, bearers, and all similar
parts of a building are made of cast iron, and it suits very
well for such purposes, because it is tough. Chimney
bars, iron ties, and other small pieces used in construction, are formed of wrought iron. Bars of fancy railing,
and balusters of stairs consist of cast iron, and, except
that they cannot resist a moderately heavy blow, no
other material is so well adapted for the purpose. These
facts will be sufficient to prove the assertion we made,
that the smith is an important personage in the art of
building, and will explain the nature of the work he has
to perform.

GENERAL REMARKS.

We have now attempted to give the reader an accurate, general notion of the several arts and trades which
are concerned in the process of building. We have

been prevented, by the space allotted to this branch of our subject, and also by the character of the volume, from entering into any particulars that did not appear to be essential to elementary knowledge, absolutely necessary for the student. A long life is hardly sufficient for the acquisition of the art of building; it may therefore be readily supposed that our greatest difficulty, in the preparation of the preceding remarks, has been to select that information which is most useful to a beginner.

PRACTICAL GEOMETRY.

PRACTICAL GEOMETRY is an important branch of knowledge to all who are in any way engaged in the art of
building. The workman, as well as the designer, requires its aid; and unless he be acquainted with some of
the leading principles of the science, he will frequently
feel an uncertainty, as to the results he may deduce from
the problems which are presented to his notice. It may
therefore be desirable, that we should attempt to explain
some of the most important of those geometrical problems
required by the builder, before we proceed to explain
the duties which devolve upon the surveyor, and the
manner in which he performs it.

PROBLEM I.

To extend at pleasure any given straight line.

. Let A B (*fig*. 43) be the straight line which it is required to extend. Take any length of the line A B as
B B, and strike C D an arc of a circle. Then with the
centre B form the intersections C and D, and, taking the
points of intersections as centres, describe the arcs in-
tersected at F. Join the points B and F, and A B F will be
a straight line; that is to say, the line A B will be continued to F in the same direction.

This problem, though very simple, is of great use to
the workman, when he has not at hand a long straight
edge or chalk-line; for, by the help of his rule and a pair
of compasses, he may draw a straight line of any length

required, since the line may be extended at pleasure. The ·workman may also, by this problem, prove the accuracy of his straight-edge ; for, having drawn the line A B F, he may fix upon any point E, and draw C D in the manner already described, and from the points C and D, as centres, describe intersecting arcs at E, as we had before done at F, and if these fall in the line that has been drawn, it will prove that the straight-edge is true.

PROBLEM II.

From any point in a given Line to erect a Perpendicular to that Line.

Let A B (*fig.* 44) be the given line, and C the point from which it is required to draw a line that shall be perpendicular to A B. On each side of the point C, take the two equal distances B D and C B. From D and B, as centres, with any radius greater than C D, describe the two arcs which cut each other in the point E. Draw the line E C, and it will be the perpendicular required.

A carpenter may, by the application of this problem, draw a line perpendicular to another, without his square, whether it be to form, upon a plank, a line square to its edge, or in other work, with his rule and compass.

PROBLEM III.

From a given Point at the End of a Line to erect a Perpendicular. ·

Let A B (*fig.* 45) be the given line ; it is required to draw from the point B, a line which shall be perpendicular to A B. From the centre B, and with any radius, describe an arc, as m n o, and, with the same radius, mark on the curve from m the point n, and from n, with the same radius, describe an arc D. Through m and n draw the line m n D, to cut the arc in D, then, through B and D, draw the line C B D, and it will be the perpendicular required.

<h2 style="text-align:center">PROBLEM IV.</h2>

To bisect a given Line.

Let A B (*fig.* 46) be the given line which it is required to bisect, or divide into two equal parts. From A and B, as centres, with any radius greater than the half of A B, describe arcs cutting each other in C and D. Draw C E D through the points of intersection, and it will bisect A B, at the point C, which was required to be done.

<h2 style="text-align:center">PROBLEM V.</h2>

Through a given Point to draw a straight Line parallel to some given straight Line.

Let A B (*fig.* 47) be the given straight line, and o the given point. From any point m in the straight line A B, describe with the radius m o, the arc o n, and from the centre o, with the same radius, describe the arc m r. Make m r equal to o n, draw the line C D through the points o and r, and it will be the line required.

<h2 style="text-align:center">PROBLEM VI.</h2>

To draw, at a given Distance, a straight Line parallel to a given straight Line.

Let A B (*fig.* 48) be the given straight line, and E the given distance. From any two points n, m, and with the radius E, describe the arcs *r* and *s*. Draw the line C D, so as to touch these arcs without cutting them, and it will be the straight line required.

PROBLEM VII.

To divide any given straight Line into any number of equal Parts.

Let A B (*fig.* 49) be the given straight line. Draw two lines, A D and B D, one at each end, parallel to each other, and set off on each, the same number of equal parts, perpendicular to A B. Join the corresponding points, and form the lines as *a e*, *b f*, *d g*, and these lines will divide A B into the equal parts 1, 2, 3, &c.

PROBLEM VIII.

To divide any given straight Line into two such Parts as shall be to each other as two given Lines are to each other.

Let A B (*fig.* 50) be the given straight line, and *m n*, and *r s*, the given proportionate lines. From the point A, draw the line A C equal to *m n* and *r s*, together, and mark upon it the line A D equal to *m n*. Join the points C and B, and draw the line D E parallel to C B : A E is to E B as *m n* is to *r s*.

PROBLEM IX.

To find a third proportional to two given Lines.

Let A B and A D (*fig.* 51) be the two given lines. So place the two lines A B and A D, that they may make any angle with each other, as at A. From A B the greater,

o

cut off a part A C, equal to A D; then join B D, and draw
C E parallel to it; A E will be the third proportional
required; that is to say, A B will be to A D as A D is
to A E.

*To form a Triangle, the Sides of which shall be equal
Lines or Lengths to be given.*

From any scale of equal parts measure the base A B
(*fig.* 52) equal to one of the given lines or lengths, then
with the centre A, and radius equal to one other of the
sides given, describe an arc, and, with the centre B and
radius equal to the third side given, describe another
arc, cutting that previously made, as in the point C. Join
A C and B C, and the sides of the triangle will be equal
to the lines given.

*The Base of a Triangle, the Perpendicular, and the
Place of the Perpendicular upon the Base being
given, to construct the Triangle.*

Lay down the base A B (*fig.* 53) by a scale of equal
parts, and mark the distance of the perpendicular from
either end of that base line, as from A, and, by a problem already given, erect the perpendicular D C, of such
a height as may be required. Join A C and B C, and the
triangle A C B will be that which is required.

. This problem will suggest to the reader the manner
in which a triangle may be measured.

PROBLEM XII.

*Upon any given straight Line to describe an equi-
lateral Triangle.*

With the radius A B (*fig.* 54), and centre B, describe
an arc; and with A as a centre, and the same radius,
describe another arc intersecting that already formed,
as in the point c. Connect A c and B c, and the triangle
A B C will be equilateral, as was required.

PROBLEM XIII.

*To bisect any Angle, that is, to divide it into two
equal Parts.*

Let A c B (*fig.* 55) be the angle which it is required to
bisect. From the centre c, and with any radius, describe
the arc *m n*. From *m*, as a centre, and with any radius,
describe an arc, and with the same radius, and *n* as a
centre, describe another arc intersecting that already
formed. Join *o*, the point of intersection, and c, and the
line uniting the two points will divide the angle A c B into
two equal parts, the angle A c *o* being equal to the angle
o c B.

PROBLEM XIV.

To find the Centre of any given Circle.

Let A D B C (*fig.* 56) be the given circle. Draw the
line or chord c D, and bisect it by the line A B in the
manner already described. Bisect the diameter A B in
the same way, by the line *m n*; the point s, in which
the lines A B and *m n* intersect each other, is the centre
required.

PROBLEM XV.

To describe a Square whose Sides shall be equal to a given Line.

Let A B (*fig.* 57) be the given line. Upon the points A and B erect the lines A C and B D perpendicular to the line given, in the manner described in Problem III. Make these perpendiculars equal to the base line A B, and join C D: A B D C is a square, the sides of which are equal to a given line.

PROBLEM XVI.

To describe a rectangular Parallelogram, the Length and Breadth of which are equal to two given Lines.

Let A B and B D (*fig.* 58) be the length and breadth given; it is required to describe a rectangular parallelogram. From the point D erect a line C D, which shall be perpendicular to B D, and make it equal to the line A B. Join A and C, and A B C D will be a rectangular parallelogram, the sides of which are equal to the lines given.

PROBLEM XVII.

To construct a regular Rhombus upon a given Line.

Let A B (*fig.* 59) be the line upon which it is required to form a regular rhombus. With the radius A B, and from A and B, as centres, describe arcs, the points of intersection being at C. Draw the line A C; and C D

parallel to A B, and B D parallel to A C; the figure A B D C
is a regular rhombus, formed upon the given line A B,
which was required to be done.

PROBLEM XVII.

*To draw a Circle which shall pass through three given
Points.*

Let A B C (*fig.* 60) be the three given points. With
any radius greater than one half A B, and from A and B,
as centres, describe arcs at G and H, and from C and B,
as centres, describe arcs intersecting each other in P
and E. Draw and continue the lines G H and E F until
they intersect each other, as in the point D; the point
of intersection is the centre of the circle required.

PROBLEM XIX.

*To draw a Tangent to a Circle from any Point in that
Circle; that is, to draw a Line from any given Point
in a Circle, in such a Manner that it shall touch the
Circumference of that Circle in the Point given,
without cutting it.*

Let B A C (*fig.* 61) be a circle, or a portion of a circle,
and A the point from which it is required to draw a tangent. From A draw A O, a radius of the circle B A C, and
from any centre, as D, draw an arc of a circle which
shall pass through A, and intersect the line A O. From
the point of intersection E, draw the line E D F, the line
intersecting the arc in the point F. Draw the line G H
through the points A and F, and the line A F H will be a
tangent to the circle A B C drawn from the given point A,
as was required.

o *

PROBLEM XX.

From any Point without a Circle, to draw a Line which shall be a Tangent to the Circle.

Let A (*fig.* 62) be a given point without the circle B C D, it is required to draw a tangent to the circle from that point. From A draw the line A o, that is, unite the point and the centre of the circle. Divide this line A o into two equal parts as at E, and with the radius E A, or E O, and from the centre E, describe the semicircle O B A, which cuts the circle in the point B. Connect the points A and B, and the line A B is a tangent to the circle B C D, drawn from the given point A, as was required.

PROBLEM XXI.

To describe an equilateral Triangle within a given Circle.

Let A B C (*fig.* 63) be a circle; it is required to draw within it a triangle whose sides are equal to one another. Commencing from any point A, mark on the circumference of the circle, a series of spaces equal to the radius of the circle, of which there will be six, and draw the arcs A D, D B, &c. Then join every alternate point as A B, B C, C A, and the several lines will together form an equilateral triangle.

PROBLEM XXII.

Within a given Circle to inscribe a Square.

Let A B C D (*fig.* 64) be the given circle, it is required to draw a square within it. Draw the diameters A B, C D, at right angles to each other; or, in other words, draw the diameter A B, and form a perpendicular bisecting it. Then join the points A C, C B, B D, D C, and the figure A B C D is a square formed within a given circle.

PROBLEM XXIII.

Within a given Circle to inscribe a regular Pentagon, that is a Polygon of five Sides.

Let A D B C (*fig.* 65) be a circle in which it is required to draw a pentagon. Draw a diameter A B, and perpendicular to it another diameter. Then divide O B into two equal parts in the point E, and join C E; and with E as a centre, and the radius C E, draw the arc C F, cutting A O in F; and with C as a centre, and the same radius, describe the arc F G; the arcs C F, G F, intersect each other in the point F, and the arc G F intersects the circumference of the circle in the point G. Join the points O and G, and that line will be a side of the Pentagon to be drawn. Mark off within the circumference the same space, and join the points A H, H I, I K, K C, and the figure that is formed, is a pentagon.

PROBLEM XXIV.

Within a given Circle to describe a regular Hexagon, that is a Polygon of six equal Sides.

Let A B C, (*fig.* 66) be the given circle, and o the centre. With the radius of the circle divide it into parts, of which there will be six, and connect the points A D, D B, &c. and the figure A D B E C F will be a regular hexagon.

PROBLEM XXV.

To Cut off the Corners of a given Square, so as to form a regular Octagon.

Let A B C D (*fig.* 67) be the given square. Draw the two diagonal lines A C, and B D, crossing each other in O: Then with the radius A O, that is half the diagonal, and with A as a centre, describe the arc E F, cutting the

sides of the square in E and F, then from B as a centre,
describe the arc G H; and in like manner from C and D
describe the arcs I K, and L M. Draw the lines L G, F I,
H M, and K E, and these with the parts of the given
square G F, I H, M K, and E L, form the octagon required.

Problem XXVI.

*To divide a given Line into any Number of Parts,
which Parts shall be in the same Proportions to each
other as the Parts of some other given Line, whether
those Parts are equal or unequal.*

Let A B (*fig.* 63) be the given line which it is required
to divide in the same manner and proportion as the line
C D, whether the parts are equal or unequal. On the
base line C D, form an equilateral triangle in the manner
already described in a former problem. Then take the
distance A B, and with E as a centre, describe the arc F G,
and join the joints F and G, and F G shall be equal to A B.
Now, if from the points H I K, which are the divisions of
the line C, we draw lines to E, as H E, I E, and K E, these
lines will cut F G, in the points *a b c*, which will divide
the line F G into parts proportionate to the divisions of
the line C D.

Problem XXVII.

*On a given Line to draw a Polygon of any Number of
Sides, so that that Line shall be one Side of a Polygon; or in other Words, to find the Centre of a Circle
which shall circumscribe any Polygon, the Length of
the Side of the Polygon being given.*

We shall here show, in a tabular form, the length of
the radius of a circle, which shall contain the given line,
as a side of the required polygon; and here we will suppose the line to be divided into one thousand equal parts,

and the radius into a certain number of like parts. The radius of the circle for different figures will be as follows:

For an inscribed Triangle 577
                    Square 701
                    Pentagon 850
                    Hexagon 1000
                    Heptagon 1152
                    Octagon 1306½
                    Nonagon 1462
                    Decagon 1618
                    Undecagon . . . 1775
                    Dodecagon . . . 1932

By this table the workman may, with a simple proportion, find the radius of a circle which shall contain a polygon, one side being given: thus, if it be required to draw a pentagon, the side given being fifteen inches, we may say as 1000 is to 15, so is 850, the tabular number for a pentagon, to 12 inches and seventy-five hundreth parts of ¯an inch, or seven tenths and a half of a tenth of an inch.

We may here give another table for the construction of polygons, one in which the radius of the circumscribing circle is given. If it be required to find the side of the inscribed polygon, the radius being one thousand parts, the sides of the different polygons will be according to the following scale:

The Triangle 1732
    Square 1414
    Pentagon 1175
    Hexagon 1000
    Heptagon 867½
    Octagon 765
    Nonagon 684
    Decagon 618
    Undecagon . . . 563½
    Dodecagon . . . 517½

Here, as in the case already mentioned, the law of proportion applies, and the statement may be thus made: as one thousand is to the number of inches contained in the radius of the given circle, so is the tabular number for the required polygon to the length of one of its sides in inches. Thus let it be supposed that we have a circle whose radius in inches is 30, and that we wish to inscribe an octagon within it; then say as 1000 is to 30 inches, so is 765 to 22 inches and 95 hundredth parts of an inch, the length of the side of the required octagon.

Having, in the preceding problems, given some elementary geometrical knowledge, we may now introduce a few remarks upon the method of drawing curved lines, bearing in mind the object of this treatise; and also give some rules for finding the forms of mouldings when they are to mitre together, that is to say, of raking mouldings, and of bevil work in general. It will also be necessary to make a few remarks upon the form of ribs for domes and groins, a knowledge of which is so necessary to the builder, that without it the workman cannot correctly execute his task. It is hardly necessary to state, that all these mechanical operations are founded upon geometrical principles, and, unless he be acquainted with these, the workman cannot hope to succeed in his attempt to excel in his art, one which is necessary for the comfort and convenience of all communities.

PROBLEM XXVIII.

To draw an Ellipse with the Rule and Compasses, the transverse and conjugate Diameters being given; that is to say, the Length and Width.

Let A B (*fig.* 69) be the transverse or longest diameter; C D the conjugate, or shortest diameter, and o the point of their intersection, that is, the centre of the ellipse.

Take the distance o c, or o d; and taking A as one point, mark that distance A E upon the line A O. Divide O E into three equal parts, and take from A F, a distance E F, equal to one of those parts. Make o g equal to o F. With the radius F G, and F and G as centres, strike arcs which shall intersect each other in the points I and H. Then draw the lines H F K, H G M, and I F L, I G N. With F as a centre, and the radius A F, describe the arc L A K, and from G as a centre with the same radius, describe the arc M B N. With the radius H C, and H as a centre, describe the arc K C M, and from the point I with the radius I D, describe the arc L D M. The figure A C B D is an ellipse, formed of four arcs of circles.

Problem XXIX.

To draw an Ellipse by Means of two concentric Circles.

Let A B (*fig.* 70) be the transverse, and E F the conjugate diameter, and o the centre of an ellipse to be drawn. From o with the radius o A, describe the circle A D B D, and from the same centre describe another circle G E H F. Divide the outer circle into any number of equal parts, the greater the number the more exact will be the ellipse and they should not be less than twelve. From each of these divisions draw lines to the centre o, as *a* o, *b* o, *c* o. Then from *a*, *b*, *c*, &c. draw lines perpendicular to A B, and from the corresponding points in the inner circle, that is, from the points marked 1, 2, 3, &c. draw lines parallel to A B. Draw a curve through the points where these lines intersect each other, and it will be an ellipse.

In the diagram to which this demonstration refers, only one quarter of the ellipse is lettered, but the process described in relation to that, must be carried round the circles, as is shown in the dotted and other lines.

PROBLEM XXX.

*To describe an Ellipse by Means of a Carpenter's
Square, or a Piece of notched Lath.*

Having drawn two lines to represent the diameters of
the ellipse required, fasten the square so that the internal angle or meeting of the blade and stock, shall be at
the centre of the ellipse.   Then take a piece of wood or
a lath, and cut it to the length of half the longest diameter, and from one end cut out a piece equal to half the
shortest diameter, and there will then be a piece remaining at one end equal to the difference of the half of
the two diameters.   Place this projecting piece of the
lath in such a manner, that it may rest against the
square, on the edge which corresponds to the two diameters; then turning it round horizontally, the two ends of
the projection will slide along the two internal edges of
the square, and if a pencil be fixed at the other end of
the lath, it will describe one quarter of an ellipse.   The
square must then be moved for the successive quarters
of the ellipse, and the whole figure will thus be easily
formed.

This method of forming an ellipse is a good substitute
for the usual plan, and the figure thus produced is more
accurate than that made by passing a pencil round a
string moving upon two pins or nails fixed in the foci,
for the string is apt to stretch and the pencil cannot be
guided with the accuracy required.

There are many other methods of drawing ellipses, or
more properly ovals, but we can only notice one or two
of those in most common use.

1. By ordinates, or lines drawn perpendicular to the

axis. Having formed the two diameters, divide the axis, or larger diameter, into any number of equal parts, and erect lines perpendicular to the several points. Next draw a semicircle, and divide its diameter into the like number of equal parts; that is to say, if the larger diameter or axis of the intended ellipse be divided into twenty equal parts, then the semicircle must be divided into the like number. As the diameter of the semicircle is equal to the shorter diameter of the ellipse, or conjugate axis, perpendiculars may be raised from these divisions of the diameter, or the semicircle, till they meet the circumference; and the different perpendiculars, which are called ordinates, may be erected like perpendiculars on the axis of the ellipse. Joining the several points together, the ellipse is described; and the more accurately the perpendiculars are formed, the more exact will be the ellipse.

2. By intersecting arches. Take any point in the axis, and with a radius equal to the distance of that point from one extremity of the axis, and with one of the foci as a centre, describe an arc; then with the distance of the assumed point in the axis from the other end of it, and with the other focus as a centre, describe another arc, intersecting the former, and the point of intersection will be a point in the ellipse. By assuming any number of points in the axis, any number of points on the curve may be found, and these united will give the ellipse. This process is founded on the property of the ellipse; that if any two lines are drawn from the foci to any point in the curve, the length of these lines added together will be a constant quantity, that is to say, always the same in the same ellipse.

P

PROBLEM XXXI.

To find the Centre and the two Axes of an Ellipse.

Let A B C D (*fig.* 71) be an ellipse, it is required to find its centre. Draw any two lines as E F and G H parallel and equal to each other. Bisect these lines as in the points I and K, and bisect I K as in L. From L, as a centre, draw a circle cutting the ellipse in four points, 1, 2, 3, 4. Now L is the centre of the ellipse. But join the points 1, 3, and 2, 4; and bisect these lines as in M and N. Draw the line M N and produce it to A and B, and it will be the transverse axis. Draw C D through L and perpendicular to A B, and it will be the conjugate or shorter axis.

PROBLEM XXXII.

To draw a flat Arch by the Intersection of Lines, having the Opening and Spring or Rise given.

Let A D B (fig. 72) be the opening, and C D its spring or rise. In the middle of A B, at D, erect a perpendicular D E, equal to twice C D, its rise; and from E draw E A and E B, and divide A E and B E into any number of equal parts, *a, b, c,* and 1, 2, 3. Join B *a*, 3 *c*, 2 *b*, and 1 A, and it will form the arch required.

The greater the number of parts into which A E and B E are to be divided the greater will be the accuracy of the curve.

Many curves may be made in the same manner, according to the position of the lines A E and E B; and if instead of two lines drawn from A and B, meeting in E, a perpendicular be erected at the same points, and two lines be then drawn from the ends of these perpendiculars, meeting in an angle, and these lines be divided into any number of equal parts, the points of the adjacent lines may be joined, and a curve will be formed resembling a Gothic arch. The demonstration already given is therefore very useful to the workman, as he may vary the form of the curve by altering the position of the lines, either with respect to the angles which they make with each other, or their proportional lengths.

PROBLEM XXXIII.

To find the Form or Curvature of a raking Moulding, that shall unite correctly with a level one.

Let A B C D (*fig.* 73) be part of the level moulding, (which we will here suppose to be an ovolo, or quarter round;) A and c, the points where the raking moulding takes its rise on the angle; F C G, the angle the raking moulding makes with the horizontal one. Draw C F at the given angle, and from A draw A E parallel to it; continue B A to H, and from C make C H perpendicular to A H. Divide C H into any number of equal parts, as 1, 2, 3, and draw lines parallel to B A, as 1 *a*, 2 *b*, 3 *c*; and then in any part of the raking moulding, as I, draw I K, perpendicular to E A, and divide I K into the same number of equal parts as H C is divided into; and draw 1 *a*, 2 *b*, 3 *c*, parallel to E A. Then transfer the distances 1 *a*, 2 *b*, 3 *c*, and a curve drawn through these points will be the form of the curve required for the raking moulding.

We have here shown the method to be employed for an ovolo, but it is just the same for any other formed moulding, as a cavetto, semirecta, &c. It may be worthy of remark, that after the moulding is worked, and the mitre is cut in the mitre-box, for the level moulding, the raking moulding must be cut, either by the means of a wedge formed to the required angle of the rake, or a box made to correspond to that angle; and if this be accurately done, the mitre will be true, and the moulding, in all its members, correspond to the level moulding. The plane in which the raking moulding is situated, is square to that of the level one ; this is always the case in a pediment, the mouldings of which correspond with the return.

Problem XXXIV.

To find the Form or Curvature of the Return in an open or broken Pediment.

Let A B C (*fig.* 74) be the angle which the pediment makes with the cornice, and let the form and size of the moulding be as in the last problem, and as shewn at D A B H. From D drop a perpendicular on C B, and draw D E perpendicular to D C, or parallel to C B;· and let D E be equal to E I, (*fig.* 73.) Then from E draw E F, parallel to D A, and divide E F into the same number of parts as I K (*fig.* 73), at 1 a, 2 b, 3 c, and transfer the distances 1 a, 2 b, 3 c, as in (*fig.* 73.) Then a curve line drawn through the points a, b, c, will be the form of the return for the moulding of the open pediment.

The mitre for the return is cut in the usual manner, but that of the pediment is cut to the proper angle of its inclination, as in the last problem. In fixing the mitre, the portion E D G of the return must be cut away, to make it come flush with the top of the pediment moulding.

THE SURVEYOR.

The business of the surveyor is to measure and value the work executed by the builder. All the departments of the art of building come under the consideration and notice of the surveyor, but, strictly speaking, he has nothing to do with the appropriateness of the work which is executed, only so far as strength and execution are concerned. Before the value of builder's work can be determined, three things must be taken into account, —the quality of the materials employed, the time and labour expended in combining them, and the quantity.

In the remarks already made, under the first division of this little volume, " The Builder," we have noticed the quality of builder's materials; but their value can only be ascertained by long practice, or by the study of such works as " The Builder's Price Book." In estimating the value of work, it is not only necessary to study the intrinsic value of the materials, but also the time and labour expended in combining them, or in suiting them to the purposes for which they are to be employed, and this cannot be calculated without practical knowledge. But there is a third element in the duties of the surveyor, and that is to ascertain the quantity of material used. This subject comes under our immediate attention ; and we shall, conseque tly, devote a few pages to a brief consideration of the fundamental facts, which may be comprised under two heads,—the measurement of superficies and solids, and

P *

the established customs in relation to the measurement of the several kinds of builder's work.

MENSURATION OF SUPERFICIES.

PROBLEM I.

To find the Area of a Square.

RULE. Multiply the side by itself, and the product will be the area.

Let it be required to find the area of a square, the side of which is 17 feet. Multiply 17 by 17, and the product will be the area; thus,

17 × 17 = 289, the area of the square in feet.

To find the side of a square, the area being given, it is only necessary to extract the square root of the area.

PROBLEM II.

To find the Area of a Rectangle.

RULE. Multiply the length by the breadth, and the product will be the area.

Thus, let it be required to find the area of a space 10 feet 7 inches, by 7 feet 3 inches. It may be found in the following manner.

	Ft.	In.	
	10	7	
	7	3	
	74	1	
	2	7	9
Feet	76	8	9

These two problems are most important in the measurement of all work that is estimated by superficies. In the measurement of painter's work, for instance, it is only necessary to take the height and the length, and to multiply the one by the other, which gives the area in square feet. There are, however, some parts of superficial work that are estimated by one dimension, that is, by the length. Although these two problems are the foundation of the art of measuring superficies, there are some others which are worthy of notice.

PROBLEM III.

To find the Area of a Rhombus or Rhomboides.

RULE. Multiply the length by the perpendicular breadth, and the product will be the area.

Let the side of a rhombus be 17 feet, and the perpendicular 15 feet, what is its area?

$17 \times 15 = 255$, the area required.

PROBLEM IV.

To find the Area of a Triangle.

RULE. Multiply the base by the perpendicular height, and half the product will be the area.

Let the base of a triangle be 14 feet, and the perpendicular height 9, then

$14 \times 9 = 126 \div 2 = 63$ the area of the triangle.

The area of a triangle may also be found from the three sides.

RULE. Add the three sides together, and from half the sum subtract each side separately; then multiply the half sum and the three remainders together, and the square root of the product will be the area required.

Let the sides of a triangle be 30, 40, and 50 feet respectively, what will be the area?

$$\frac{30 \times 40 \times 50}{2} = \frac{120}{2} = 60, \text{ half the sum of the sides.}$$

$$60 - 50 = 10, \text{ first remainder.}$$
$$60 - 40 = 20, \text{ second remainder.}$$
$$60 - 30 = 30, \text{ third remainder.}$$

then $60 \times 10 \times 20 \times 30 = 360000.$

$\sqrt{360000} = 600$ the area in feet.

PROBLEM V.

Any two Sides of a right-angled Triangle being given, to find the third Side.

1. When the base and perpendicular are given.

RULE. To the square of the base add the square of the perpendicular, and the square root of the sum will give the hypothenuse, or longest side.

Let the base of a right-angled triangle be 24, and the perpendicular 18, what is the hypothenuse?

576 square of the base
324 square of the perpendicular.
$$576 \times 324 = 900$$
$\sqrt{900} = 30$, the hypothenuse.

2. When the hypothenuse and one side are given.

RULE. Multiply the sum of the hypothenuse and one side by their difference; the square root of the product will give the other side.

If the hypothenuse of a right-angled triangle be 30, the perpendicular 18, what will be the base.

30 + 18 = 48 sum of the two sides.

30 — 18 = 12 difference of the two sides.

48 × 12 = 576

$\sqrt{576}$ = 24 the length of the base.

PROBLEM VI.

To find the Area of a Trapezium.

RULE. Divide the trapezium into two triangles, by a diagonal drawn from one angle of the figure to its opposite. The areas of the triangles may be found by the rules already given, and the sum will give the area of the trapezium. It is unnecessary to give an example of this problem, as it would be only a repetition of what has been already illustrated.

To find the area, of irregular polygons, or many-sided figures, it is only necessary to reduce them into triangles, and parallelograms, and, calculating these severally, to add them together: the sum will give the area of the figure. In this manner the land surveyor estimates the quantity of acres, roods, and perches, contained within certain boundaries; and it may be done with considerable accuracy, by subdividing the space until the whole area is contained within a number of single figures. The surveyor of builder's work, however, has seldom a necessity for this mode of proceeding; for it is customary, in all those cases where a surface has a variable height, to take the medium between the two extremes, and consider the superficies as a parallelogram. But as the builder is sometimes required by circumstances to measure the ground which is chosen as the site of a building, it is necessary that he should be able to do so when required ; and we doubt if the dismemberment which the profession has of late suffered, and the ignorance of professional men upon certain subjects, is at all advantageous to either themselves or their clients.

Problem VII.

Either the Diameter or Circumference of a Circle being given, to find the other.

1. To find the circumference, the diameter being given.

Rule. As 7 is to 22, so is the diameter to the circumference.

If the diameter of a circle be 84·5 inches, what is the circumference ?

As 7 : 22 : : 84·5 : 265·571.

Therefore 265·571 is the circumference required.

2. To find the diameter, the circumference being given.

Rule. As 22 is to 7, so is the circumference to the diameter.*

Problem VIII.

To find the Area of a Circle.

1. When the diameter and circumference are both given.

Rule. Multiply half the circumference by half the diameter, and the product will be the area.

2. When the diameter is given.

Rule. Multiply the square of the diameter by ·7854, and the product will be the area.

---

* The above rules are not strictly correct. The following is a much nearer approximation. Multiply the diameter by 3.1416, for the circumference ; and, divide the circumference by 3.1416, for the diameter.

3. When the circumference is given.

RULE. Multiply the square of the circumference by ·07958, and the product will be the area.

We might add many other useful problems to those already demonstrated, but we must refer the reader to those works which are written on the subject, for further information. Mensuration is a branch of science which cannot be considered unimportant to the architectural student, or the young workman, though it is often passed over by the one as well as by the other. It is true that a very slight knowledge of one or two problems is sufficient to enable the surveyor to measure the superficies of builders' work; but we should not form a very high estimate of the capacity and energy of a mind, that was ready to receive any dogma that might be stated as a fact, without inquiring into the cause or origin of the fact itself. And yet there are hundreds who use rules for no other reason than that they were used by their predecessors; and they themselves have been taught by their predecessors to employ them. The amount of knowledge, and the variety of learning, required of the architect, may be pleaded as an excuse for ignorance; but the very difficulties and the labour required ought to stimulate to exertion. It is very disgraceful to be ignorant of that which most men understand; but there is no honour in possessing the information. There is no honour in knowing as much and doing as much as other men; but there is honour in knowing and doing more. Every individual, whatever be his station and engagements, ought to attempt superiority, and to raise himself, if possible, above the rank of society in which he happens to be born. Poverty is not a bar to advancement, but a stimulus to exertion; and it is a singular fact, that those who have signalized themselves in their particular pursuits, have generally advanced, by personal exertion, from the stations to which they had a sort of hereditary claim. We have been induced, by the consideration of the subject more immediately under our attention, to make these remarks, that we may prompt the student's ambition. We may be permitted to say to every reader, and especially to those who have recently devoted themselves to the study of the art of building, read and thoroughly understand the six first books of Euclid. Putting out of consideration the application of this knowledge to the purposes of the builder and architect, we may state that the man who has sufficient energy of mind to attempt and fully accomplish this task, is a hopeful man, for he shows decision of character. We should prefer, as a friendly adviser, an individual who had thoroughly mastered the six books of Euclid's Geometry, to the light superficial students and reputed scholars of the present day. It is scarcely possible to read so much of Euclid, without feeling that the mind has had a useful mental discipline; and if no other advantage were gained, this would repay the student for the time which he expended over the study. The habit of close application, and of tracing effects to their cause, cannot be obtained without some trouble; but the quality of mind induced will have a great influence upon all the future engagements of life. A man of close application, accustomed to inquire into the cause of the state or being of principles and things, will not be easily led into the gross errors which frequently destroy weaker minds.

We may now proceed to an explanation of a few problems in the art of measuring solids. We shall not however, introduce many, but confine our attention to those which appear most useful to the surveyor, or which can be applied in other departments of the art of building.

MEASUREMENT OF SOLIDS.

Solids are those bodies which have length, breadth, and thickness; and they are distinguished from each other by their figure; thus, a cube, a prism, a parallelopipedon, a cylinder, a pyramid, a sphere. The measurements of all these are founded on the principles of geometry; and as the surveyor may have occasion to use all the rules, in practice, it is necessary that he should be perfectly acquainted with them, and with the principles from which the rules are deduced.

PROBLEM I.

To find the Solidity of a Cube.

DEFINITION. A cube is a solid enclosed by six equal square surfaces.

RULE. Multiply the side of the square by itself, and that product by the side of the square, and the product of the two multiplications will give the solidity of the cube.

If the side of a cube be 9 feet, what is its solidity?

$$9 \times 9 = 81$$
$$81 \times 9 = 729, \text{ the solidity required.}$$

PROBLEM II.

To find the Solidity of a Parallelopipedon.

DEFINITION. A parallelopipedon is a solid, having six sides, every opposite two being equal and parallel to each other. *Fig.* 75.

Q

RULE. Multiply the length by the breadth, and the product by the depth, and it will give the solidity required.

If the length of a parallelopipedon be 82 inches, its breadth 54, and its depth 10, what is its solidity?

$$82 \times 54 = 4428$$
$$4428 \times 10 = 44280, \text{ the solidity required.}$$

PROBLEM III.

To find the Solidity of a Prism.

DEFINITION. A prism is a solid, the ends of which are parallel, equal, and of the same figure. Specific names are given to them, according to the form of their bases or ends : if they are triangles, the prism is called a triangular prism; if rectangles, rectangular, and so on. *Fig.* 76.

RULE. Multiply the area of the base by the perpendicular height, and the product will be the solidity required.

What is the solidity of a rectangular prism whose base is 30 inches and height 53?

$$30 \times 53 = 1590, \text{ the solidity in inches.}$$

PROBLEM IV.

To find the Solidity of a Cylinder.

DEFINITION. A cylinder is a round prism, having circles for its ends, and is formed by the revolution of a right line about the circumference of two equal circles, parallel to each other.

RULE. Multiply the area of the base by the perpendicular height of the cylinder, and it will give the solidity.

PROBLEM V.

To find the Solidity of a Sphere.

DEFINITION. A sphere is a solid formed by the revolution of a semicircle round a fixed diameter.

RULE. Multiply the cube of the diameter by ·5236, and the product will be the solidity.

It must not be supposed that we consider the problems, introduced into this little volume, all that are necessary for the person who is engaged in the measurement of builder's work. The limits of an introductory treatise prevent us from introducing more than those we have already given; and we have chosen them because they appeared to us the most important. We may therefore now advance to a consideration of the methods adopted in the measurement of work.

MEASUREMENT OF BRICKLAYER'S WORK.

The bricklayer generally undertakes the digging for foundations, and the execution of all kinds of brickwork, tiling, and sometimes slating.

Digging for Foundations. There are two methods of estimating the value of excavating: it may be done by allowing so much a day for every man's work, or so much per cubic yard for all that is excavated, the price being, of course, regulated according to the difficulties attending the process, and the distance to which the earth that is dug out must be wheeled. And here it may be necessary again to remark, that our object is not to give the prices of materials or labour, but to explain the manner in which it is measured; that is to say, the process by whlch the dimensions are found.

We have already explained the method of finding the solidity of a cube, and of a parallelopipedon; and it is therefore hardly necessary to state, that to find the cubical quantity in a trench, or an excavated area, the length, width, and depth must be multiplied together. These are usually given in feet, and therefore, to reduce the amount into cubic yards, it must be divided by 27.

Let us suppose a trench 40 feet long, 3 feet deep, and 3 feet wide. These dimensions will give, as the product, 13 yards 9 feet; and will stand in the following manner in the surveyor's book:

Ft.	In.	Yds.	Ft.
40	0		
3	0	13	9
3	0		

Measurement of Brickwork.

Brickwork is measured by the rod, which is 272¼ feet square; but it is customary to reduce all work to 1½ brick thick, and consequently the rule given for the measurement of excavations does not exactly apply in this case.

RULE. Multiply the number of superficial feet contained in the wall by the number of half-bricks which the wall is in thickness, and divide the product by three, that is, three half-bricks. Then divide the product by 272, and that will reduce it to rods.*

* It is customary in practice to reject the quarter foot in the division by the number of feet in a rod.

Let it be required to find the number of rods of reduced brickwork contained in a wall 50 feet long, 10 feet high, and two bricks thick.

$$
\begin{array}{r}
50 \\
10 \\
\hline
500 \\
\end{array}
$$

4 No half-bricks in thickness.

$$
3)\ 2000
$$

$$
272)\ 666\ 8
$$

$\cdot\cdot$ 2 rods. 122 ft. 8 in.

All openings (such as for doors and windows,) are deducted from the amount, the whole face of brickwork being measured as though there were no openings.

Measurement of Chimneys.

When a chimney does not adjoin a party-wall, it may be measured taking the girt in the middle, the height of the story, and the depth of the jambs, since the thickness must be equal to the depth of the jambs.

A chimney shaft is measured by taking the girt, and the height; something being allowed for the trouble and expense of scaffolding, according to circumstances.

Tiling and Slating.

Tiling and slating are measured by the square of 100 feet, double measure being sometimes allowed for hips and valleys. Sky-lights and chimney-shafts are deducted.

Q *

How many squares of tiling are there in a roof of 27 feet, 6 inches deep, on both sides, and 50 feet long?

$$
\begin{array}{rr}
\text{Ft.} & \text{In.} \\
50 & 0 \\
27 & 6 \\
\hline
25 & 0 \\
350 & \\
100 & \\
\hline
100) \quad 1375 & 0 \\
\hline
\end{array}
$$

·· 13 squares, 75 feet.

MEASUREMENT OF CARPENTER AND JOINER'S WORK.

To measure carpenter and joiner's work with accuracy, not only requires an intimate acquaintance with the method of executing the work, but also with the peculiarities of custom, which no ingenuity can deduce from any general principle, and for which no reason can be given. It may, however, be taken as a general rule, that all timbers are cubed, and boards measured superficially, but mouldings are valued by their length.

It may be desirable, before we make any particular remarks, to introduce a table, which may serve the place of a series of definitions.

Cubic Feet.

A load of rough timber 40
A load of squared timber . . . 50
A cord of wood . . . . 128 (8 feet long, 4 feet broad, and 4 feet deep.)

Cubic Feet.

A stack of wood . . . . 108 (12 feet long,
3 feet broad, and 3 feet deep.)
A ton of shipping 42
A floor of earth 324
A solid yard of earth, 1 load.

All joists, girders, and in fact all the parts of naked
flooring, are measured by the cube, and their quantities
are found by multiplying the length by the breadth, and
the product by the depth. The same rule applies to the
measurement of all the timbers of a roof, and also the
framed timbers used in the construction of partitions.

The following table may be useful to the student.

Cubic Inches.	Cubic Foot.	Cubic Yard.
1728	1	
46656	27	1
7762392	4492½	166⅝
496793088000	287496000	10648000
254358061056000	14719795200	5451776000

Flooring, (that is to say, the boards which cover the
naked flooring,) is measured by the square. The dimensions are taken from wall to wall, and the product is divided by 100, which gives the number of squares; but de-
ductions must be made for staircases and chimneys.

Boarded partitions are measured in the same manner
as flooring, deduction being made for doors and windows,
except otherwise agreed upon. Weather-boarding is
also sometimes measured by the square, though some
surveyors prefer to take it by the yard square.

Doorcases, staircases, wainscotting, doors, and shut-
ters, are measured by the foot superficial. Windows are
valued by the foot superficial, or, as is sometimes preferred, at a sum per window.

Beads, fillets, skirtings, boxings to windows, and such work, is generally valued by linear measure.

MEASUREMENT OF MASON'S WORK.

The method of measuring mason's work does not much differ from that adopted in measuring carpenter and joiner's work; that is to say, some parts are measured by the foot cube, some by the foot superficial, and some by the length. Blocks, pillars, columns, and such like, are valued by the foot cube, and generally all work that is more than two inches thick. Slabs, chimney-pieces, and pavior's work, is estimated by the foot superficial; mouldings and other small work by the foot run.

There is much difficulty in the measurement of mason's work, more in fact than in any other branch of building. This arises from the difference of opinion among surveyors, as to the most equitable method of measuring. Such differences of opinion ought to be settled, for when we consider the value of the material and the labour in many cases, it is important to the public that some fixed rules of practice should be adopted.

MEASUREMENT OF PLASTERER'S WORK.

Plasterer's work is in part measured by the yard and the foot square, and in part by the foot run. Rendering, lath and plastering, stucco, and pugging, are valued by the yard; floated friezes and soffits by the foot superficial; cornices and mouldings by the foot run. In rendering, no deduction is made but for doors and windows, except in rendering between the quarters, that is, where the braces and timbers project beyond the plastering; one-fifth is then generally deducted. Whitening and colouring are measured in the same manner as plastering. All enrichments are taken singly, and valued according to their size and the richness of their workmanship. If there be more than four angles in a room, they are allowed for. All cornices and mouldings, and all works where the running mould is used, are measured from the nose of the moulding to the wall, and we speak of a moulding as being so many inches according to its girt.

MEASUREMENT OF PAINTER'S WORK.

Nearly all painter's work is measured by the square yard, and all is measured over which the brush passes. Cornices, mouldings, narrow skirtings, reveals to doors and windows, and generally all work, not more than nine inches wide, are valued by their length. Sash-frames are charged so much each, according to their size, and the squares so much a dozen. Mouldings cut in, are charged by the foot run, and the workman always receives an extra price for party-colours. Writing is charged by the inch, and the price given is regulated by the skill and manner in which the work is executed; the same is true of imitations and marbling. The price of painting varies exceedingly, some colours being more expensive and requiring much more labour than others.

In measuring open railing, it is customary to take it as flat work, which pays for the extra labour; and as the rails are painted on all sides the two surfaces are taken. It is customary to allow all edges and sinkings.

MEASUREMENT OF PLUMBER'S WORK.

All pipes are charged by linear measure, according to their thickness and diameters; sheet lead is estimated per cwt. Nearly all other articles are charged, each according to the weight of lead it contains, and the trouble of making. To these prices the workman's time must be added.

MEASUREMENT OF GLAZIER'S WORK.

Glaziers take their dimensions in feet, inches, and tenths, and estimate by the foot square; for windows, it is only necessary to take the dimensions of one pane, and multiply the product by the number of panes. Sometimes, however, the surveyor will measure the length and breadth of the window, including bars. In fancy-work, the greatest length and breadth are taken, to compensate for the loss of material and the labour required.

Having made these few remarks upon the method of measuring builder's work, we may possibly assist the student by giving him a few pages from a surveyor's book, though practice alone can give him all the information he requires.

BRICKLAYER'S WORK.

ft.	in.	Reduced.		
81	4	ft.	in.	
	9	101	8	2½ brick footings.
81	9			
17	6	1430	7	1½ brick external wall.
81	4			
1	10	99	5	1 brick parapet wall.

ft.	in.	Reduced.		
5	4	ft.	in.	
8	9	31	1	1 brick chimney breast.
3	0			
3	6	10	6	1½ brick, deduct opening.
3	6			
4	6	26	3	2½ brick chimney shaft.
3	0			
4	9	14	3	1½ brick, deduct window.
4	0			
6	0	24	0	1½ brick, deduct window.
6	0			
7	0	28	0	1 brick wall of strong closet.
3	0			
4	0	8	0	1 brick arch to ditto.
	3	4		
2	1	6	10	0	Gauged arches.
	5	10		
2	1	6	17	6	Ditto.

CARPENTER'S WORK.

23	4			
10	6	245	0	2½ square, labour and nails to common span-roof with collar, struts and ceiling joists.
72	8			
	5			
	4	10	1	Fir wall plate.

	ft.	in.	Reduced.		
	6	0	ft.	in.	
		5			
		4		10	Add laps.
		—			
	16	4			
3		7	10	9	Tie beams.
		4½			
		—			
	72	8			
		4	6	0	Pole plate.
		3			
		—			
	13	3			
		9	1	3	Ridge.
		1½			
		—			
	10	0			
4		9	3	9	Hips.
		1½			
		—			
	14	6			
8		4	8	0	Rafters.
		2½			
		—			
	23	4			
2		7	6	350	0	¼ battening for slates.
		—			
	78	8			
		1	8	131	1	1¼ deal gutter and bearers.
		—			
	22	0			
13		3½	13	11	Ceiling joists.
		2			
		—			

JOINER'S WORK.

ft.	in.	ft.	in.	
6	9			
2	9	18	7	1½-inch deal 6-panelled door, moulded on both sides.
16	6			
	4	5	6	1¾-inch deal rebated and beaded jamb linings.
16	6			
	5	11	0	Inch square grounds.
16	10	33	8	OG moulding.
6	9			
3	0	20	3	2-in deal 6-panelled door, bead flush and square.
2	6			
3	6	17	6	2-in deal ovolo sash hung folding.
12	4			
	3	3	1	¼-in. mitred and beaded lining.
3	0			
5	6	16	6	1½-in. moulded and square shutters.
171	3			
	6	85	8	Inch deal wrought and beaded linings.
5	0			
2	0	10	0	Inch wainscot counter-top, on deal framed brackets.

R

ft. in.	ft. in.	

5 0
2| 9   7 6   1¼-in. deal shelves and brackets.
_____

2 9
10| 1 3   34 4   1½-in. deal steps and risers.

8| 2 9   22 0   Inch·square bar-balusters.

MASON'S WORK.

7 0
7 0   49 6   York paving.
_____

6 6
1 2   7 7   York steps.
_____

    Four mortice holes.

33 7   33 7   Bath proper sunk and throated sills.
_____

.86 0   86 0   12-inch feather-edged Bath coping.
_____

5 4
4| 1 6   32 0   1½-in. Portland slab.
_____

11 0
4| 6   22 0   Mantle and jambs.
_____

4 4
4| 7   9 1   1½-in. shelf.
_____

    Two rounded corners, four times.
    Two notches to slabs, four times.

3 0
4| 9   9 0   York inner hearth.
_____

ft.	in.	ft.	in.	
5	6			
1	9	9	8	2½-in. Portland cover over entrance door.
4	6			
4	6	20	3	6-in. rubbed York landing.
17	0			
	7	4	2	Portland jambs, and head to door-way.
	5			
7	0			
7	0	49	0	Thin rubbed York paving, bedded in cement, and cramped to walls.

PLASTERER'S WORK.

ft.	in.	ft.	in.	
15	7			
15	4	238	11	Lath, plaster, float, sett, and white ceiling.
5	4			
	9	4	0	Deduct chimney breast.
7	0			
7	0	49	0	Add second room.
3	0			
1	6	4	6	Deduct angle.
14	0			
2	6	35	0	Add strings of stairs.
63	4			
7	3	459	2	Trowelled stucco for paint, to walls.

	ft.	in.	ft.	in.	
	25	0			
	7	3	181	3	Add to staircase.
	6	3			
	5	0	31	3	Deduct window.
	3	9			
	5	0	18	9	Ditto.
	6	9			
2	3	0	40	6	Deduct doors.
	3	0			
2	3	6	21	0	Deduct chimney opening.
	5	6			
	2	6	13	9	Ceiling under bow-window.

PLUMBER'S WORK.

	82	8			
	12	11	1167	9	6 lb. milled lead to gutters.
	82	8			
		6	41	4	4 lb. milled lead for flashings.
	53	3			
	2	6	133	2	5 lb. milled lead for hips and ridges.
2	30	0	60	0	4-in. iron rain-water pipe.
					2 cistern heads.

ft.	in.	ft.	in.	
6	0			
1	0	9	0	6 lb. milled lead for sink.
3	0			
1	0	3	0	Ditto.
4	0	4	0	1½-in. waste pipe to sink.
				1 chain and plug to sink.
4	6			
4	0	18	0	5 lb. milled lead over bow.

PAINTER'S WORK.

Three oils, grained wainscot, and varnished.

ft.	in.	ft.	in.	
7	1			
4	0	28	4	Door.
16	6	16	6	Lining.
44	6			
8	0	356	0	Partitions.
59	4	59	4	Torus skirting.

Two window frames.
Three dozen squares.

ft.	in.	ft.	in.	
3	0			
4	9	14	3	Shutters.

Four oils.

ft.	in.	ft.	in.	
63	4			
7	3	459	2	Partitions and walls.
25	0			
7	3	181	3	Add for Staircase.

R*

ft. in.	ft. in.	
5 0		
1 7	7 11	Deal shelves and brackets.
8 2 9	22 0	In square bar balusters.
11 0	11 0	Deal framed newel.
12 6	12 6	Deal moulded hand-rail.
8 0		
3 4	26 8	Closet front.
52 10	52 10	Chair-rail.

SLATER'S WORK.

23 3		
15 0	348 9	3½ squares best countess slating.
9 6		
4 1 0	38 0	Cuttings to hips.

GLAZIER'S WORK.

1 2		
28 1 0	32 8	Best seconds glass.
1 6		
12 1 1	19 6	Best crown glass.

SMITH'S WORK.

20 feet plain iron railing.

4 cast-iron gratings to arbour windows, let into stone jambs.

33 feet wrought iron railing to stairs.

3 pair 3-inch brass iron-butt hinges
and screws.

3, 7-inch iron-rim locks and screws.
Iron cramps and plugs for mason
  and plumber.
4, 6-inch iron bolts.
1 iron-barrel chain.

We have now explained the manner in which the surveyor enters his measurements, as well as the manner in which the several works are measured. For the value of the materials and workmanship, we must again refer the reader to The Builder's Price Book. We have, however, another task to perform, before we can pass on to the third part of our little book, and that is to introduce a few tables, which may be useful to the student, as the basis of calculations that he may under several circumstances be required to perform.

Table of the Cohesive Strength of Bodies.

	Pounds Avoirdupois.
Iron rod an inch square will bear	76,400
Brass	35,600
Hempen rope	19,600
Ivory	15,700
Oak, box, yew, plumtree	7,850
Elm, ash, beech	6,070
Walnut, plum	5,360
Red fir, holly, elder, crab	5,000
Cherry, hazel	4,760
Alder, asp, birch, willow	4,290
Lead	430
Freestone	914

This table is the result of a series of experiments made by the celebrated Emerson; but Mr. Barlow, speaking

of the results, says, "they all fall very short of the ultimate strength of the woods to which they refer."

Mr. George Rennie made some experiments upon the
resistance of timbers to crushing, the following results
were obtained.

English oak, (base one inch, square length one inch,)
    crushed by 3860
White deal 1928
American Pine 1606
Elm 1284

The following table gives the relative strength of different woods, the beams being supported on one end.
Each kind being 4 feet long, 2 inches broad, and 2
inches deep.

Kind of Wood.	Weight in Pounds that broke the Piece.
English Oak . . .	260
Dantzic Oak . .	210
Riga Fir	210
Pitch Pine . . .	270

In the following results, pieces 3 feet long, 2 inches
broad, and 2 inches deep were used:

	lbs.
Beech	401
Ash	436

In the following results, pieces 5 feet long, 2 inches
broad, and 2 inches deep:

	lbs.
Green ash . . .	239
Teak	257
Virginian yellow pine .	147
Canadian white pine .	132
Dry larch	162

·It is sometimes necessary to determine the specific gravity of bodies; that is, to find the proportional weight of bodies, in relation to some standard. The whole theory of sinking and floating should be thoroughly understood by the student. A balloon does not rise in the atmosphere, a cork float upon water, or lead sink in it, without a cause; but, when any substance sinks in a fluid, it is because it is heavier than an equal bulk of that fluid; when it floats, it is because it is lighter.

The builder does not often require a table of specific gravities, but it may be sometimes serviceable, and we shall select those substances which he most frequently employs.

Table of Specific Gravity and Weight of Woods.

	Specific Gravity.	Weight of a Cubic Foot Avoir. lbs.
Poplar	·383	23·94
Larch	·544	34·00
Elm	·556	34·75
Honduras Mahogany .	·560	35·00
Poom · . . .	·579	36·18
Willow	·585	36·56
Cedar . . .	·596	37·25
Pitch pine . . .	·660	41·25
Pear-tree . . .	·661	41·31
Walnut . . .	·671	41·94
Mar forest tree . .	·694	43·37
Elder-tree . . .	·695	43·44
Beech	·696	43·50
Orange-wood . .	·705	44·06
Cherry-tree . . .	·715	44·68
Teak	·745	46.56

	Specific Gravity.	Weight of a Cubic Foot Avoir. lbs.
Maple and Riga fir	·750	46·87
Ash and Dan. oak	·760	47·50
Yew, Dutch . . .	·788	49·25
Apple-tree . .	·793	49.56
Alder	·800	50·00
Yew, Spanish . .	·807	50·44
Mahogany, Spanish .	·852	53·25
Oak, Canadian . .	·872	54·50
Box, French . . .	·912	57·00
Logwood . . .	·913	57·06
Oak, English . .	·970	
Ebony . . .	1·331	83·18
Lignum Vitæ . .	1·333	83·31

The following Table may be very useful, as giving the weight of several materials commonly used in building.

14·835 cubic feet of	Paving stone	weight 1 ton.
14·222	—	Common stone	—
13·505	—	Granite	—
13·070	—	Marble	—
12·874	—	Chalk	—
11·273	—	Limestone	—
64·460	—	Elm	—
64·000	—	Honduras mahogany	—
51·650	—	Mar forest fir	—
51·494	—	Beech	—
47·762	—	Riga fir	—
47·158	—	Ash and Dantzic oak	—
42·066	—	Spanish mahogany	—
36·205	—	English oak	—

THE ARCHITECT.

ARCHITECTURE has been sometimes defined the art of building; but it is more properly the art of designing buildings, according to those principles which civilized man has acknowledged to constitute beauty, and to those which science has proved to be necessary for stability. If architecture be the art of building, it may be found among the savages of Africa and Australia, as well as in the civilized societies of Europe and America;—the man who constructs a rude mis-shapen hut, as his only shelter from the turbulence of the elements, has as much right to be called an architect, as the masters of antiquity who designed the magnificent temples of Greece and Rome. But architecture is not the art of building; it is the art of designing according to established laws, calculated to secure strength, comfort, and beauty.

We are not ignorant that there are many persons who very highly esteem the architecture of the present day, and flatter themselves that some of the buildings erected in our own age may go down to posterity as the models of perfection. Where those specimens of taste are to be found, we know not; but why they are not to be found, is quite evident. The blame is not to be attached to the architect, for there are men capable of executing works as far superior to the bald and prison-like structures which haunt the metropolis, as they are, in strength, to the mud-hut of the Indian. Unfair, or rather mock, competition, is the parent of that darkness

which must envelope the architectural talent of the
country. In the present day, influential friends are
more valuable than talent; and there has never been a
period in the history of the world, in which the architect
has been less esteemed as a man of science, learning, and
taste. How differently the architects of antiquity were
encouraged and treated, may be gathered from the celebrated letter of the Emperor Theodosius to Symmachus.

"The dispositions of our palace are so well ordered,
that our learned artists cannot pay too much to conserve
it, since the admirable beauty of this chief work, if not
kept in repair, in the end would be destroyed by the
lapse of time. These excellent constructions make my
delights; they are the noble image of the power of the
empire, and they attest the greatness and glory of kingdoms. The palace of the monarch is represented to the
ambassadors as an edifice worthy of their admiration;
and, at first sight, the master appears to them such as
his habitation seems to announce. It is then a great
pleasure to a prince, who is a connoisseur, to inhabit a
palace which unites all the perfections of the art; and
there refresh his mind from the occupation of public
affairs, by the charm which the marvels of his edifice
procure him. It is said that the Cyclops were the first
who built, in Sicily, edifices as spacious as the caverns
which they had abandoned, after Ulysses had deprived
the unfortunate Polyphemus of sight. It was from
thence that the art of constructing passed into Italy;
and posterity, rivals of these first architects, profited
from their inventions, and employed them for their
necessities and comfort.

"From this we notify, that your intelligence and
talents have determined us to confide to you the care of
our palace. Our desire is, that you be attentive to pre-

serve, in its ancient splendour, all that is antique, and
that what you add, be constructed in the same taste;
for, as a beautiful form ought to be clothed with a
uniform colour, in like manner it is befitting that the
same beauty and the same taste reign in all the members
and parts of our palace. By often reading Euclid, and
imprinting on your mind the astonishing variety of
figures with which he has enriched his books of geometry,
you will be rendered capable of accomplishing our in-
tentions, and be in immediate possession of matter to
answer our requests. Have also always in view the
profound lessons of Archimedes and Metrobes, in order
to enable you to produce new works of merit. This is
not an employment of little consequence which is confided to you, since it obliges you to accomplish, by the
ministry of your art, the ardent desire we have to illus-
trate our reign with new edifices. For, whether we wish
to repair a city, build fortresses, or yield to the flattering
pleasure of erecting a prætorium, you will be obliged to
execute, and give a sensible existence, to the objects on
which we may determine. What employment more
honourable, what office more glorious, than this, which
places you within the reach of transmitting, to the most
distant ages, edifices which will ensure you the admiration of posterity! For you are required to direct the
mason, sculptor of marble, founder of bronze, workmen
in stucco and plaster, and painter in mosaic. You are
bound to teach them that of which they are ignorant,
and to resolve the difficulties which this army of men,
who work under your guidance, and who are to have
recourse to your enlightened judgment, propose to you.
Behold, then, how much he ought to have, who has so
many to instruct. But you will gather the fruits of
their labours; and the success of their works, which you

shall have well conducted, will make your eulogy, and
will become your most flattering recompense. For this
reason we wish, whatever you may be charged to build,
that it be done with so much intelligence and solidity
that the new erections may only differ from the ancient
in the freshness of their date. This will be possible to
you, if a base cupidity never incline you to deprive the
workman of a part of our bounty. It is easy to make
yourself obeyed, if they receive an honest and competent salary, without fraud or reserve. A generous hand
animates the genius of the arts; and all the ardour of
the artist is directed to his work, when he is not dis-
tracted by care for a subsistence. Further, consider
what the distinctions are with which you are decorated :
you walk immediately before our person, in the midst
of a numerous retinue, having the golden rod in hand,
a prerogative which, by your approaching so near to us,
announces that it is to you that we have confided the
execution of our palace."

From this letter we learn that architecture was highly
esteemed by the ancients, and that the highest honours
were offered to those who successfully pursued the
practice. The mercenary motives with which the profession is followed in the present day, arise in a great
measure from the withdrawal of those inducements to
excellence, which were offered to the ancients, and even
to our forefathers. The student soon discovers that
honour is not to be immediately obtained, even by pre-
eminent talent, and that there is but one thing to be
procured by the pursuit of his profession, and that is
money. But wealth is not always the attendant of great
skill and honourable exertion; mediocrity of talent,
barefaced impudence, and cunning, too frequently secure
it. A knowledge of this, and it is a fact, however un-

pleasant it may be to acknowledge it, is often found sufficient to suppress that enthusiasm by which alone eminence can be attained, and to excite to those practices which are inimical to the dignity of the profession. The architect should however remember, that the public taste is somewhat under his control, and that, although undeviating honour, and an attention to his own dignity, may tend, for a period to withhold success, yet time, the corrector of all things, will bring him to due honour.

These remarks lead us to the inquiry, what is required of the architect? or rather, what should be his character, talents, and attainments? We leave Vitruvius to answer this question, and, as the writings of this celebrated author may not be in the possession of all our readers, we shall give a translation of the chapter "on the Education of an Architect."

"An architect should be ingenious and apt in the acquisition of knowledge. Deficient in either of these qualities, he cannot be a perfect master. He should be a good writer, a skilful draughtsman, versed in geometry and optics, expert in figures, acquainted with history, informed on the principles of natural and moral philosophy, somewhat of a musician, not ignorant of the sciences both of law and physic, nor of the motions, laws, and relations to each other, of the heavenly bodies. By means of the first named acquirements, he is to commit to writing his observations and experience, in order to assist his memory. Drawing is employed in representing the forms of his designs. Geometry affords much aid to the architect: to it he owes the use of the right line and circle, the level and the square, whereby his delineations of buildings on plane surfaces are greatly facilitated. The science of optics enables him to introduce with judgment the requisite quantity of light according to the aspect. Arithmetic estimates the cost, and aids in the measurement of the works; this, assisted by the laws of geometry, determines those abstruse questions, wherein the different proportions of some parts to others are involved. Unless acquainted with history, he will be unable to account for the use of many ornaments which he may have occasion to introduce. For instance, should any one wish for information on the origin of those draped matronly figures, crowned with a mutulus and cornice, called Caryatides, he will explain it by the following history. Carya, a city of Peloponnesus, joined the Persians in their war against the Greeks. These, in return for the treachery, after having freed themselves, by a most glorious victory, from the intended Persian yoke, unanimously resolved to levy war against the Caryans. Carya was, in consequence, taken and destroyed; its male population extinguished, and its matrons carried into slavery. That these circumstances might be better remembered, and the nature of the triumph perpetuated, the victors represented them draped, and apparently suffering under the burthen with which they were loaded, to expiate the crime of their native city. Thus, in their edifices, did the ancient architects, by the use of the statues, hand down to posterity a memorial of the crime of the Caryans. Again, a small number of Lacedæmonians, under the command of Pausanias, the son of Cleombrotus, overthrew the prodigious army of the Persians at the battle of Platæa. After a triumphal exhibition of the spoil and booty, the proceeds of the valour and devotion of the victors were applied by the government in the erection of the Persian portico; and as an appropriate monument of the victory, and a trophy for the admiration of posterity, its roof was supported by statues of

the barbarians, in their magnificent costume; indicating,
at the same time, the merited contempt due to their
haughty projects, intimidating their enemies by fear of
their courage, and acting as a stimulus to their fellow-
countrymen to be always in readiness for the defence of
the nation. This is the origin of the Persian order for
the support of an entablature; an invention which has
enriched many a design with the singular variety it exhibits. Many other matters of history have a connexion
with architecture, and prove the necessity of its professors being well versed in it.

"Moral philosophy will teach the architect to be
above meanness in his dealings, and to avoid arrogance:
it will make him just, compliant, and faithful, to his
employer; and, what is of the highest importance, it will
prevent avarice gaining an ascendency over him; for he
should not be occupied with the thoughts of filling his
coffers, nor with the desire of grasping every thing in
the shape of gain; but, by the gravity of his manners
and a good character, should be careful to preserve his
dignity. In these respects we see the importance of
moral philosophy; for such are her precepts. That
branch of philosophy which the Greeks called φυσιολογία,
or the doctrine of physics, is necessary to him in the
solution of various problems; as, for instance, in the
conduct of water, whose natural force, in its meandering
and expansion over flat countries, is often such as to
require restraints which none know how to employ but
those who are acquainted with the laws of nature; nor
indeed, unless grounded on the first principles of physic,
can he study with profit the works of Ctesibius, Archimedes, and many other authors who have written on the
subject. Music assists him in the use of harmonic and
mathematical proportion. It is, moreover, absolutely

s *

necessary in adjusting the force of the balistæ, catapultæ, and scorpions, in whose frames are holes for the passage of the homotona, which are strained by gut-ropes attached to windlasses worked by hand-spikes. Unless these ropes are equally extended, which only a nice ear can discover by their sound when struck, the bent arms of the engine do not give an equal impetus when disengaged, and the strings, therefore, not being in equal states of tension, prevent the direct flight of the weapon. So the vessels called ηχεια by the Greeks, which are placed in certain recesses under the seats of the theatres, are fixed and arranged with a due regard to the laws of harmony and physics, their tones being fourths, fifths, and octaves; so that when the voice of the actor is in unison with the pitch of these instruments, its power is increased and mellowed by impinging thereon. He would moreover be at a loss in constructing hydraulic and other engines, if ignorant of music. Skill in physic enables him to ascertain the salubrity of different tracts of country, and to determine the variations of climates, which the Greeks called κλιματα: for the air and water of different situations being matters of the highest importance, no building will be healthy without attention to those points. Law should be an object of his study, especially those parts of it which relate to party walls, to the free course and discharge of the eaves waters, the regulations of cesspools and sewage, and those relating to window-lights. The laws of sewage require his particular attention, that he may prevent his employers being involved in law-suits when the building is finished. Contracts, also, for the execution of the works, should be drawn with care and precision; because when without legal flaws, neither party will be able to take advantage of the other. Astronomy instructs him in the points of the heavens, the laws of the celestial bodies,

the equinoxes, solstices, and courses of the stars; all of
which should be well understood in the construction and
proportions of clocks. Since, therefore, this art is founded
upon and adorned with so many different sciences, I am
of opinion that those who have not, from their earliest
youth, gradually climbed up to the summit, cannot, without presumption, call themselves masters of it. Perhaps,
to the uninformed, it may appear unaccountable that a
man should be able to retain in his memory such a variety
of learning; but the close alliance with each other, of the
different branches of science, will explain the difficulty;
for as a body is composed of various concordant members,
so does the whole circle of learning consist in one harmo-
nious system. Wherefore those who from an early age
are initiated in the different branches of learning, have a
facility in acquiring some knowledge of all, from their
common connexion with each other. On this account
Pythius, one of the ancients, architect of the noble temple
of Minerva at Priene, says, in his commentaries, that an
architect should have that perfect knowledge of each art
and science which is not even acquired by the professors
of any one in particular, who have had every opportunity
of improving themselves in it. This, however, cannot
be necessary; for how can it be expected that an Architect should equal Aristarchus as a grammarian, yet should
he not be ignorant of grammar? In music, though it be
evident he need not equal Aristoxenus, yet he should
know something of it. Though he need not excel, as
Apelles, in painting, nor as Myron or Polycletus, in
sculpture, yet he should have attained some proficiency
in these arts. So, in the science of medicine, it is not
required that he should equal Hippocrates. Thus also, in
other sciences, it is not important that pre-eminence in
each be gained; but he must not, however, be ignorant of

the general principles of each. For, in such a variety of matters, it cannot be supposed that the same person can arrive at excellence in each, since to be aware of their several niceties and bearings, cannot fall within his power. We see how few of those who profess a particular art arrive at perfection in it, so as to distinguish themselves: hence, if but few of those practising an individual art, obtain lasting fame, how should the architect, who is required to have a knowledge of so many, be deficient in none of them, and even excel those who have professed any one exclusively. Wherefore Pythius seems to have been in error, forgetting that art consists in practice and theory. Theory is common to and may be known by all, the result of practice occurs to the artist in his own art only. The physician and musician are each obliged to have some regard to the beating of the pulse, and the motion of the feet; but who would apply to the latter to heal a wound or cure a malady? so, without the aid of the former, the musician affects the ears of his audience by modulations upon his instrument. The astronomer and the musician delight in similar proportions, for the position of the stars, which are quartile and trine, answer to a fourth and fifth harmony. The same analogy holds in that branch of geometry which the Greeks call λογος οπτικος: indeed, throughout the whole range of art there are many incidents common to all. Practice alone can lead to excellence in any one; that architect, therefore, is sufficiently educated, whose general knowledge enables him to give his opinion on any branch when required to do so. Those unto whom nature has been so bountiful that they are at once geometricians, astronomers, musicians, and skilled in many other arts, go beyond what is required of the architect, and may be properly called mathematicians, in the extended sense of the word. Men so gifted, discriminate accurately, and are rarely met with. Such, however, was Aristarchus of Samos, Philolaus and Archytas of Tarentum, Apollonius of Perga, Eratosthenes of Cyrene, Archimedes and Sopinas of Syracuse : each of whom wrote on all the sciences."

It would be considered invidious to compare the present state of architecture with what it was when such varied acquirements as those stated by Vitruvius were considered necessary. But we cannot too strongly impress this consideration upon the student, for it will incite him to great activity and a determination to use his best endeavours for success. Plato says, that a good architect was a rarity in Greece ; it need not be so in Britain,—we have talent enough in the country for all other pursuits, why not for this ? We and our forefathers have formed too low an opinion of the knowledge, both in the extent and the objects, required of those who pursue this most noble profession. We do not fear to state that architecture requires a more varied and profound knowledge both of literature and science than any other pursuit; and yet it is usually supposed that a knowledge of drawing is all that is required. This is the source of all the mischief; it has lowered the profession in the estimation of observing and intelligent men, and it has done more, it has dismembered the profession, and created party jealousies that are highly unbecoming and injurious. The civil engineer complains of the architect, because he is not capable of executing works which require an acquaintance with scientific and mathematical principles; the architect considers the engineer to be a man without taste, and the surveyor charges them both with an ignorance of the method of measuring and valuing the works they design. Let the reader apply

these remarks as they are intended by the author, not as
a charge against those who are now practising the art;
for there are many, very many, who combine an extensive learning with an integrity of principle that does
honour to the profession, as well as to themselves; but
as a reason why students should devote themselves,
with a more than common earnestness, towards the
acquisition of those several branches of knowledge
which are necessary for a successful practice.

In the remarks we are about to make upon architecture, it will be our object to direct the attention of the
student to some of those principles, which are acknowledged, on all hands, to be the constituents of beauty,
and to adduce a few examples in which these principles
have been successfully employed. These objects will
be best secured by a brief history of the art, with such
details, in relation to particular structures, as may appear
necessary.

There is more philosophy in the old proverb, "necessity is the mother of invention," than is usually
admitted. Those arts and sciences which yield most
advantage to man in his social and individual capacity,
have been the first to advance towards perfection.
Man insensibly surrounds himself with comforts, which
give birth to luxuries. This is consistent with his
nature. It cannot he denied that indolence and personal
gratification are elements in the constitution of the
human mind; and if this position be admitted, it accounts for the acknowledged fact that the sciences of
practical utility have first arrived at maturity. But,
although we make this assertion, we do not forget that
even in the days of Homer but few, if any, of the now
acknowledged principles of architecture had been determined. The fact, however, rather supports than

opposes our theory. Poetry is a beautiful representation of a vivid imagination, and architecture is the application of imagination directed by utility to a specific purpose— the comforts and luxuries of life. The arts, generally, and architecture especially, as soon as its principles were determined, have either kept pace with or preceded poetry : they are guided by the same faculty, and when one has declined the other has participated in the influence that produced the decrease. So, also, as the one advanced, it attracted the other. But poetry is the representation of the imaginative faculty, conveyed to the ear, and transmitted to posterity by oral tradition or by a written language, without any other faculties than those in the possession of the poet ; whereas architecture, on the other hand, requires not only an imagination, but wealth, to give a material existence to the ideas of the designer. The poet may present to the mind of the observer a beautiful picture, but the architect gives being to his representation ; and, to 'siqı op something more than imagination is required.

If we were to trace the art of architecture back to its origin, we should tire the reader with a long disquisition upon primitive huts, as many before us have done ; and little advantage could, in the end, be obtained from all our researches. We shall therefore content ourselves, and we hope our readers, by commencing with the earliest recorded period of architectural skill.

Syrian Architecture.

Babylon is one of the most celebrated cities of antiquity, but it is doubtful whether it owes its origin to Nimrod or to Semiramis. All antiquaries, however, admit, that it was a city of immense extent, and that the tower of Babel, mentioned in the Holy Scriptures, as the cause of the confusion of tongues, was contained within its walls.   This building is supposed to be the same as that called the Temple of Belus, which; according to Strabo, consisted of eight square towers, rising one above the other, and connected with a general staircase on the outside.

This city was afterwards enlarged and embellished by Nebuchadnezzar, the personage whose history is given in Scripture as a warning to the presumptuous.   The city was bounded by a wall of immense thickness, surrounded by a ditch; and a hundred gates of brass, defended by towers, led to its interior.

Palmyra was another city of importance in Syria, and is supposed by some antiquaries to have been built by Solomon.   It is now a mass of ruins, but it is magnificent even in its decay, and presents many objects of interest to the classical traveller.

Balbec, or Heliopolis, is celebrated for a beautiful temple of the sun, supposed to have been erected by Antoninus Pius.

Nineveh, a city celebrated in ancient history, is stated to have been of immense extent, and to have contained many buildings of great magnitude.   We learn from history that its walls were three hundred feet high, and that three chariots could have been driven abreast upon them.

Persian Architecture.

If we turn from that part of the world first inhabited by man, to other districts early possessed by him, we shall find the same evidence of the extent and splendor of ancient architecture.   But it is worthy of remark, that the magnificence of these ancient cities did not consist in the elegance or in the accurate proportions of the buildings, but in their vastness, and the rich and costly materials employed in their construction. This is true, not only of Persia but also of India, Egypt, and the countries which preceded them. Persepolis was, so far as history informs us, the principal city in ancient Persia, and we have an evidence of the statement in the extent and grandeur of the ruins. There are, however, no ruins of temples, for the opinions of the ancient Persians prevented their erection. They were Tsabaists, and held the doctrine that it was improper to worship in places where the Deity would be confined by walls; under the supposition that a being who could fill the universe could not dwell in a temple made with hands.

Persepolis was celebrated for a magnificent palace, called by the inhabitants *Chehul Minar*, or *Tsehil Minar*, the place of forty columns. The style has a great resemblance to that of the Egyptian. The blocks of stone are of huge dimensions, and bear inscriptions in Arabic, Persian, and Greek; many of these have, according to Dr. Hyde, been written in honour of Alexander the Great. The whole structure was composed of deep, grey, hard marble, susceptible of a fine polish.

Indian Architecture.

There was a great peculiarity of character in all that was done by the Hindoos, and they seem to have advanced above all the ancient nations in knowledge generally, and especially in the sciences of quantity and number. It is a singular circumstance, and worthy of notice, that this people had acquired so extensive an acquaintance with the exact sciences, that they dis-

T

covered and used the Binomial theorem and other mathematical rules, supposed to be of modern invention, until the translation of Sanscrit manuscripts, by Sir William Jones, and other oriental scholars. The principal specimens of Indian architecture which have descended to us, are vast excavations that have been apparently used as temples. That in the island of Elephanta is supposed to be the most ancient, and next to it those in the islands of Ellora, Salsette, and Canarah. But we must not look at these as if they were the only records of the art and science of the Hindoos, for there is abundant evidence that in later times they erected some very beautiful temples; among which, that of Benares was probably pre-eminent; for a column, which is still in existence, is considered the finest specimen of eastern art.

The temple in Elephanta is a square, of about 135 feet on each side, and 14½ feet high. The roof is supported by ranges of columns, which are more elegant than those of the Egyptians, and the walls are covered with gigantic figures, sculptured in relievo. The excavations of Salsette are near the village of Ambola. The temple is a square of about 28 feet, and is approached by a long walk; at the end of which there is a gateway, 20 feet high, leading to a grand vestibule. The roof of the temple is supported by twenty columns, resembling those at Elephanta, and about 14 feet high; but as they are composed of a softer stone, they are not so well preserved.

Egyptian Architecture.

There is a difference of opinion among antiquaries and architects as to the relative merits of the Hindoos and Egyptians. Some evidence may certainly be adduced to prove that the Hindoos were the fathers of science and art, and that the Egyptians derived more assistance from them than they were willing to acknowledge. But, however this may be, it cannot be denied that there is a general resemblance between the two styles of architecture; but we know much more of the character of the Egyptian than of the Indian.

It is not necessary that we should, in this general sketch, enter into any particulars in relation to the remaining specimens of Egyptian architecture. It was distinguished by its massiveness and the richness of its ornaments. Memphis was one of the principal cities of Egypt, and is said to have been built by Menes, A.C. 2188. Near to this city stood the celebrated pyramids, the largest of which covered a space of 435,600 square feet; and the Sphinx Ghizà was in the same neighbourhood. All the sacred buildings of Egypt were decorated with massive columns, generally without a base, the capitals being of various descriptions; sometimes consisting of a single abacus, but having commonly a bell shape, reversed, and variously decorated. The walls were thick and built of stones, embellished with emblematic devices, which the moderns have been able to interpret. The roofs also were formed of stone in immense blocks, and wide steps of the same material were commonly provided at the entrance of the temple, between sphinxes of enormous size.

Egyptian architecture is solemn, and frequently sepulchral, characterized by a stiffness of contour, solidity, and massiveness. But, at the same time, it is the parent of all that airy elegance and grace which so peculiarly distinguishes the Grecian. The walls of Egyptian buildings were usually thick, and the roof consisted of a single stone, supported by pillars of different shapes, round, square, or octagonal. It has been stated that the Egyptians were unacquainted with the arch, but Dr. Pocock objects to this assertion, and Belzoni found rude specimens at Thebes and Gournou. The principal Egyptian buildings were the pyramids, obelisks, labyrinths, monolithal chambers, sphinxes, and temples.

The largest of the pyramids is situated a few miles from Cairo, and each side of its base, which is a square, is said to be 660 feet; its height is about 500 feet, and its summit finishes with a platform about 16 feet square. It is a singular circumstance that authors greatly differ in their measurements, as will be seen by the following table.

Authors.	Height.	Width of one Side.
Herodotus	800 feet.	800 feet.
Strabo	626	600
Diodorus	600	700
Pliny	—	708
Thevenot	520	612
Le Brun	616	704
Niebuhr	440	710
Greaves	444	648

But we pass from the review of these comparatively speculative subjects, to one which immediately directs the practice of the art in the present day. Grecian architecture is in itself so perfect, that no improvement

has been made in its details, from the period when it
was at its highest state among its inventors, until the
present moment; a circumstance which proves more
than any other the excellence of the principles they
adopted.

Grecian Architecture.

It is impossible to determine the period when architecture received the elements of perfection in Greece.[*]
We are, in fact, left in so much ignorance of the subject,
that we do not even know whether it was the result of
the attention given to it in a single age, or whether it
was of slow growth, advancing with the civilization of
the people. History is silent as to the infancy of this
noble science; and we examine it, first of all, in the
majesty and beauty of its maturity.

Grecian architecture consists of three orders; the
Doric, the Ionic, and the Corinthian. The essential
parts of an order are a column and an entablature; and
the orders are distinguished from each other by the forms
of the bases and capitals of the columns, as well as by
the height and form of the shaft, and the details of the
entablature. There are eight mouldings introduced in
the orders: the ovolo, the talon, the cyma, the cavetto,
the torus, the astragal, the scotia, and the fillet. But
these are not used promiscuously, each one has its place,
and the position in one order does not warrant the con-
clusion that it has necessarily the same place in the
others. We shall, however, now proceed to an explanation of the orders separately, and to give such particular information as our limits will permit.

* It may perhaps be referred most properly to the time of Pericles,

T *

The Doric Order.

The Doric is admitted to be the oldest of all the Grecian orders of architecture. Its origin and its name are
involved in as much obscurity as its advance to perfection. According to some writers, it was invented by
Dorus, king of Achaia and Peloponnesus, the son of
Helena; and was employed in the temple built by him
in honour of Juno, at Argos. There seems, however, to
be some evidence that it was known and adopted by the
Dorians, before it came into general use among the
Greeks; and therefore other persons have stated that it
derives its name from this circumstance. That it was
introduced into Greece at a very early period in the history of that country, there can be no doubt, for Vitruvius,
having no better evidence, was compelled to adopt a
fable to account for its invention. "Dorus, king of
Achaia, and all Peloponnesus, the son of Helenus, and
of the nymph Optice, having once caused a temple to be
built to Juno, in the ancient city of Argos, this temple
was of the style we call Doric. Afterwards this order
was employed in all the cities of Achaia, without having
as yet any established rule for the proportions of its architecture. But as they, the Greeks, were unacquainted
with the proportions which it was necessary to give to
columns, they sought the means of making them sufficiently strong to sustain the weight of the edifice, and
to render them agreeable to the view. For that end they
took the measure of the foot of a man, which is the
sixth part of his height, after which measure, they
formed their columns in such a manner that, in proportion to this measure, which they gave to the thick-

ness of the foot of the column, they made it six times that height, including the capital : and thus the Doric column, which was first employed in the edifices, had the proportion, force, and beauty, of the body of a man."

This statement is extremely absurd in itself, and is opposed to the common progress of the arts and sciences. No really valuable information is obtained by guessing, and no science is born in a state of maturity. Perfection can be only obtained after a long and continued application of the mind to the pursuit. It was thus with the Grecian Doric order, for at first it was rude and heavy, but by successive improvements its proportions were at last determined, and it assumed a light though stable appearance. The temple of Corinth is an example of the early Doric, and it will be observed, when comparing this specimen with others, that the column is shorter, and the capital less projecting than in more recent examples.

The Grecian Doric column has no base, not even tori or fillets. The shaft is sometimes fluted, and sometimes plain; and we find a few examples in which it is fluted on the top and bottom only. The flutes are wide and not sunk to any great depth.

The following are the principal existing specimens of Grecian Doric: the temple at Corinth; the temples of Theseus and Minerva; the Parthenon; the temple of Jupiter Nemæus, between Argos and Corinth; the Propylæa and Portico of the Agora at Athens; the temple of Apollo in the island of Delos; the temples of Juno Lucina and Concord at Agrigentum; the temple at Egesta; and those at Pæstum and Silenus.

For further particulars concerning this order, we would direct the attention of the reader to Mr. Aikin's work on the subject. The following remarks on the antiquity of the Doric, and the early introduction of fluted columns, are very curious. Other orders are, in fact, only distinguished by their capitals; but the Doric order bears its characteristic marks in every part: the shaft and the entablature are not less peculiar than the capital; and this circumstance, independently of external evidence, appears to me to prove this order not only to be the primitive and original architecture, but to have its composition founded upon fixed principles, and some acknowledged type.

It is remarkable that the works of Homer, whose genius was so observant, and whose style is so circumstantial, should afford so little information as to the state of architecture in his time; for, although he sometimes mentions temples of the gods, he never describes them. Indeed, all his ideas of architectural beauty seem confined to the barbaric magnificence of precious materials. Thus the palace of Menelaus and Alcinöus are inlaid or encrusted with ivory, brass, and silver; but scarcely a hint is given of their construction. We, however, learn chiefly from the example of the palace of Ulysses, which is made the scene of so many transactions that some description of its form necessarily occurs, that, at this early period, the great halls of houses had columns placed internally. These columns were rather for use than ornament, and were fixed for support of the roof; for they are mentioned in connexion with the beams and rafters. One circumstance, however, seems to shew that even these early and domestic examples were fluted, which is a considerable mark of the genuine Doric order, when compared with the architecture of other countries. This occurs in the first book of the Odyssey, where Telemachus, receiving as a guest, Minerva, in the form of Mentes, places her spear within a spear-holder in the column, which must be, in all probability, a channel or flute.

It will not be expected that, in a little volume, like this, which the author is now presenting to the student, any allusion should be made to the method of drawing the orders; and it is the less necessary, because Mr. Nicholson's excellent work on the Five Orders, is, or ought to be, in the hands of every person who studies the art of building. But it may be useful to the student if we give him, in a tabular form, the details of one or two of the best specimens in every order.

		Temple of Theseus at Athens.			Temple of Minerva at Athens.		
		Height of the Members.			Height of the Members.		
		Modules.	Parts.	Fractions.	Modules.	Parts.	Fractions.
Entablature {	Cornice ..	—	25	½	—	26	—
	Frieze ...	1	25	—	1	19	—
	Architrave	1	20	—	1	14	½
Column {	Capital....	1	—	—	—	28	—
	Shaft	10	—	—	10	2	—
Height of the Column ..		15	11	—	—	15	—

The Grecian Ionic Order.

The Ionic order differs in so many respects from the Doric, that its invention is one of the most noble productions of a fertile imagination; and the proportions are so true and elegant, that it is scarcely possible to imagine by what process of thought so beautiful a form should be suggested. It is stated by historians, that it was invented by Hermogenes of Alabanda, and was employed by him in the erection of a temple to Bacchus. The inventor being a native of Caria, then possessed by the Ionians, the order was called the Ionic.

The volutes of the capital may be considered the distinguishing feature of this order; but, in almost all other respects, it differs from the Doric in its proportions, forms, and in the adoption of a base. The most celebrated examples of this order are the temples of Ilissus, Erectheus, and Minerva Polias, at Athens; the temple of Bacchus, at Teos; the temple of Apollo Didymæus, at Miletus; the aqueduct of Hadrian, at Athens; and the temple of Minerva Polias, at Priene.

The Ionic has always been a favourite order among the nations who have been acquainted with it. The simple elegance of its column, the gracefulness of the capital, and the proportions of the entablature, give it a character which no other style can excel. It has also another peculiarity; its outline is elegant, and, if left in its simple grandeur, it presents a graceful character; but it will also admit of ornament, as a beautiful form may be decorated with a splendid dress; in either case it has charms. But, although we speak thus highly of the Ionic, we cannot but blame the exclusive use of it. In England, nearly all the public buildings have Ionic porticos. In almost every street of the metropolis we may find an imitation of an Ionic temple. This exclusive attention to a particular style must have a tendency to cramp the imagination of the architect, and make him a mere copyist. It is true that we may find no forms so elegant as those transmitted to us by the ancients; and it is quite certain that if the Greeks had made no greater exertions than ourselves, we should not have had them. As well might the poet cease to write because he cannot surpass Homer, as the architect to design because he cannot invent a more perfect form than the Ionic.

		Temple of Ilissus at Athens.			Temple of Minerva Polias at Athens.		
		Height of the Members.			Height of the Members.		
		m.	*p.*	*f.*	*m.*	*p.*	*f.*
Entablature	Cornice ..	1	2	—	1	7	¼
	Frieze ...	1	19	—	1	18	¼
	Architrave	1	25	—	1	21	¾
Column	Capital ...	—	27	¼	1	13	—
	Shaft	14	2	½	16	22	¾
	Base	1	—	—	—	24	¼
Height of the Column ..		20	16	—	23	17	¼
Volute		1	6	—	1	5	¼

The Corinthian Order.

The Greeks have the honour of introducing the Corinthian, as well as the other orders of architecture to which we have already alluded. Vitruvius accounts for the origin of the Corinthian by the following tale, which is usually supposed to be a fable.

"A young lady, at Corinth, fell ill and died. After her burial, her nurse collected together sundry ornaments, with which she used to be pleased; and putting them into a basket, placed it near her tomb; and, lest they should be injured by the weather, she covered the basket with a tile. It happened that the basket was placed on the root of an acanthus, which, in springing, shot forth its leaves; these running up the side of the basket, naturally formed a kind of volute, in the turn given to the leaves by the tile. Happily Callimachus, a most ingenious sculptor, passing that way, was struck with the beauty, elegance, and novelty of the basket, surrounded by the acanthus leaves; and, according to this idea or example, he afterwards made columns for the Corinthians, ordaining the proportions, such as constitute the Corinthian order."

It has been maintained by some writers, that the Corinthian capital is but an elegant modification of the Egyptian, and that the Greeks invented it from the study of Egyptian art. The celebrated M. Quatremere de Quincy has expressed his belief in this supposition. It is argued by these writers that Greece was at first but an Egyptian colony, for Athens was founded by Cecrops, an Egyptian. It is also well known that all the early Grecian philosophers and artists repaired to Egypt, for the purpose of studying the sciences and arts of that ancient country. There is, in fact, a great resemblance between the general elements of the Egyptian and the Corinthian capital, though the one is heavy and destitute of proportion, the other light and elegant.

The Corinthians were always celebrated as the patrons of art. They were essentially a commercial people, and, consequently, became wealthy. But, having enriched themselves, they indulged in all the luxuries that could be obtained, embellished their city with temples, theatres, and palaces, and engaged the most celebrated artists of the period to design and execute them. Corinth was, therefore, not only the most opulent city in Greece, a character given to it by Thucydides, but also the most elegant.

Unfortunately, the hand of time and a desolating war has destroyed every specimen of Corinthian architecture, in the noble city whose name it bears; and, indeed, but few specimens remain in any part of Greece.

The following Table gives the Proportions of the three most important remains of Corinthian Architecture in Greece.

		Choragic Monument of Lysicrates.			Arch of Theseus at Athens.			Temple of Jupiter Olympus at Athens.		
		Height of the Members.			Height of the Members.			Height of the Members.		
		m.	p.	f.	m.	p.	f.	m.	p.	f.
Entablature	Cornice .	1	20	—	1	16	—	1	18	
	Frieze ..	1	9	¼	1	10	—	—	21	½
	Architrave	1	21	—	1	12	¼	1	11	¾
Column	Capital .	2	23	—	2	15	—	2	7	—
	Shaft ..	16	16	—	15	5	—	16	7	—
	Base ...	21	21	—	1	10	—	1	1	—

The great superiority of the Greeks, in the noble art of architecture, did not arise from any one cause, but resulted from a variety of circumstances, none of which could alone have raised this people to so high an approximation to perfection. It must, however, be mainly attributed to the warm and energetic imagination of the Greeks, aided by external circumstances; such as the freedom which they enjoyed, and the exuberant beauty of the country they called their own. In many other countries both law and religion restrained the energies of the people, and especially prevented the progress of the arts; but, in Greece, both lent their aid, and, however painful may be the reflection, it cannot be doubted that the wonderful advance of the science of architecture in that country must be, in part, traced to the splendid colouring with which the mythological notions of the people were decorated, and the influence it had upon the public mind.

It is impossible to discover the precise era in which

the arts were introduced into Greece; and, if we could do so, we should find, judging from the ordinary progress of the arts, that, at that period they were as destitute of those beauties which afterwards distinguished their progress, as were other ancient nations. This opinion is greatly strengthened by the circumstance, that, in the age which produced the works known as the writings of Homer, and they must have been written at a period nearly contemporaneous with the return of the Heraclidæ, no determined principles or proportions had been introduced into the Grecian architecture; for the poet does not dwell upon the grace and elegance of the structures, but upon the splendour and value of the materials with which they were constructed.

It has been very confidently stated by some writers, that the science of architecture was brought into Greece by Cadmus, who lived about fifteen hundred years before the Christian era, and is said to have founded the city of Thebes, so called after the Egyptian city of the same name. However this may be, we have the evidence of Tacitus, that its ruins were extensive and very magnificent.

But little advance in architecture appears to have been made, for a long period, in any of the Grecian kingdoms; for it was probably in the colonies of Asia Minor that the Doric and Ionic orders were invented. The Corinthian was not produced until many years afterwards, when the arts had advanced to almost perfection. This, the richest of all the orders, was invented in Greece properly so called.

> " First unadorned
> The nobly plain, the manly Doric rose ;
> Th' Ionic then, with decent matron grace
> Her airy pillar heaved ; luxuriant, last,
> The rich Corinthian spread her wanton wreath,

The whole so measured true, so lessened off
By fine proportion, that the marble pile,
Formed to repel the still or stormy waste
Of rolling ages, light as fabrics looked
That from the magic wand aerial rise.
These were the wonders that illumined Greece
From end to end."

The different styles of Grecian architecture warrant, according to the opinions of some writers, the division of the history of the art into five periods; and, although we may not perfectly agree with those authors in opinion, it will not be disadvantageous to the reader if we take a general review of the data upon which they found their theory, before we advance to a consideration of the state of the art among the Romans.

The first is the Homeric period, which includes the works of Trophonius, who built the temple of Delphos, Agamedes, and Dædalus. During this era, architecture was not regulated by any fixed or defined proportions; but it was probably in a state of incipient prosperity, advancing towards that perfection which afterwards distinguished it. Homer, speaking of the palace of Priam, says that it had fifty apartments at its entrance, inhabited by the princes and their wives; and that they were surrounded with porticoes. We have other evidence to prove that columns were employed at this early period, but the character of the architecture cannot be ascertained.

The second period is that from Rhœcus of Samos, who lived about seven hundred years before Christ, to the time of Pericles; and during this Ctesiphon, Callimachus, and other celebrated architects, flourished.

The third period commenced with Pericles and closed with Alexander the Great. This was the most brilliant era of Greece, that in which Hippodamus of Miletus,

Phidias, Calicrates, and other celebrated artists, flourished. During this period, architecture was at its highest elevation; the orders had received the last touches of the master minds of Greece, and the nation itself retired from the field of glorious victory to breathe the atmosphere of peace, and to enjoy the productions of genius, both in poetry and the arts.

The fourth period commenced with the death of Alexander the Great, and continued till the time of Augustus. Dinorates, Sostrates, and Taurus, are the most celebrated architects of this period. Vitruvius is supposed to have flourished in the time of Augustus.

The fifth era continued from the reign of Augustus till Constantinople was made the seat of government; and then Grecian architecture fell.

ROMAN ARCHITECTURE.

For all that was excellent in the architecture of the Romans, this people were indebted to the Greeks; but, if we could give this great nation a negative praise without qualification, we should be happy to do so; unfortunately, there is much truth in a statement which has often been made :—that they deteriorated all that they attempted to improve. The Grecian orders were adopted in Rome, but were so altered as to lose all the dignity and grace by which they had been characterised. The Roman architecture is remarkable for redundancy of ornament, massiveness and splendour, and for extent; but never for grace or elegance. The Roman architects were never excelled by any other nation, in the scientific arrangements adopted in construction, or the skill in providing for the comfort which ought always to be studied in designing a building. In the description which it is necessary to give of the style of the Romans, we shall first of all give a table of the proportions adopted in the most celebrated of the Roman buildings, and then describe the orders which were peculiarly their own, adding a few general remarks.

Roman Doric Order.

The following is a Table of the Height of the Members of the principal Specimens of Roman Doric.

		Theatre of Marcellus at Rome.			*Thermæ of Dioclesian at Rome.*		
		Height of the Members.			Height of the Members.		
		m.	*p.*	*f.*	*m.*	*p.*	*f.*
Entablature	Cornice . .	1	6		1	16	—
	Frieze . . .	1	16	—	1	15	—
	Architrave	1	1	0	1	2	—
Column	Capital . . .	1	2	$\frac{1}{3}$	1	—	—
	Shaft	14	15	$\frac{2}{3}$	15	—	—

Roman Ionic Order.

		Temple of Fortuna Virilis at Rome.			*Theatre of Marcellus at Rome.*			*Colosseum at Rome.*		
		Height of the Members.			Height of the Members.			Height of the Members.		
		m.	*p.*	*f.*	*m.*	*p.*	*f.*	*m.*	*p.*	*f.*
Entablature	Cornice .	2	10	$\frac{1}{4}$	2	6	—	1	19	—
	Frieze . .	—	28	$\frac{3}{4}$	1	6	$\frac{2}{3}$	1	16	$\frac{1}{2}$
	Architrave	1	8	$\frac{1}{2}$	1	13	—	1	10	$\frac{1}{3}$
Column	Capital .	—	21	$\frac{1}{2}$	—	19	$\frac{1}{2}$	—	18	$\frac{1}{3}$
	Shaft . . .	15	10	—	16	10	$\frac{1}{3}$	15	23	$\frac{1}{4}$
	Base . . .	1	—	$\frac{3}{4}$	1	—	—	—	27	$\frac{1}{2}$
Volute		—	29	—	—	25	$\frac{1}{2}$	—	24	—
Pedestal	Cornice .	—	23	$\frac{1}{4}$	—	19	$\frac{1}{2}$	2	17	—
	Die . . .	3	3	$\frac{3}{4}$	3	12	$\frac{1}{4}$	—	20	$\frac{1}{2}$
	Base . . .	2	3	$\frac{3}{4}$	—	—	—	1	13	—

Roman Corinthian Order.

		Temple of Jupiter Stator at Rome.			Temple of Jupiter Tonans at Rome.			Portico of the Pantheon at Rome.		
		Height of the Members.			Height of the Members.			Height of the Members.		
		m.	p.	f.	m.	p.	f.	m.	p.	f
Entablature	Cornice .	2	9	½	1	16	¼	1	24	—
	Frieze . .	1	13	½	1	13	—	1	9	½
	Architrave	1	13	½	1	8	—	1	12	¾
Column	Capital .	2	6	½	2	10	—	2	7	½
	Shaft . .	17	—	—	17	4	½	16	8	½
	Base . . .	1	—	—	1	—	—	1	—	—

These three tables will enable the student to compare the proportions adopted in Rome with those invented by the Greeks; and he may find it useful to compare the general character of the buildings by the examination of well authenticated drawings, or, if possible, details, bearing the several dimensions in his memory. But we may now pass on to make a few remarks upon the two orders which were invented by the Romans, the Composite and the Tuscan.

The Roman Tuscan Order.

The Tuscan style of architecture has been almost universally considered as a separate order, but is, in fact, only a variation of the Doric, with some alterations in the proportions. All that we know of this singular modification of the Doric is that derived from Vitruvius, who has given the proportions. The diameter of these columns, he says, taken at the bottom, should be a seventh part of their height, and their height should be a third of the breadth of the temple. Their diminution at the top should be a fourth of their diameter at the bottom. The base should be half a diameter in height, and composed of a circular socle or plinth, having a height equal to half the base; and a torus, which, with its fillet, should be the same height as the plinth.

This order is seldom employed in the present day, and was not probably extensively used even by the Romans. The finest specimen in this country, and perhaps in the world, is the church of St. Paul's, Covent Garden, designed and executed by Inigo Jones. It has been said, with truth, by a modern writer, that this order, with its great projection of the crown members, over the long cantilivers, may be applied with the greatest propriety to market-places.

The Roman Composite Order.

If we meant by an order of architecture a particularly formed capital, then that which is called the composite order would be properly designated. But as it is equally unwise to form such an idea of the meaning of an order, as it would be to suppose the variations in the countenance of men to distinguish them in species, so it was highly improper to give so much importance to a peculiar alteration of the Corinthian capital, as to designate it an order. The term, however, may now be continued, and can do no harm, provided that the student understand the limited sense in which it is used.

The Composite column differs from the Corinthian no more than the various specimens of Corinthian do among themselves. The Composite capital seems to be a union of the capitals of the Corinthian and Ionic. The Composite order appears to have been first used in the triumphal arches which the Romans were accus-

tomed to erect to commemorate great victories. There
are many fine remains of this style, but the most celebrated is the arch of Titus, erected in honour of that
prince to commemorate his conquest of Jerusalem.

From these remarks, it will be evident, that there are
in fact but three orders of architecture; the Doric, which
has the general character of strength; the Ionic, which
is the representation of elegance and grace; and the
Corinthian, which has the quality of richness. The
Romans, as we have already seen, possessed these, as
well as the Greeks; yet the character of the architecture
of these two nations was as different as can be imagined.
The imagination of the Greeks, though vivid, was chas-·
tened, and distinguished by elegance and propriety.
The mind of the Roman was more pleased with show,
and vastness of design, than with grace; and was besides proud, stern, and unbending. If history were
silent, we might suppose this from a knowledge of their
architecture, which is ornamented, massive, and usually
the specimens are of great extent.

Having presented the reader with tables showing the
height of the members, in all the orders of architecture,
as employed by both Greeks and Romans, we shall now
add three other tables, which will give the projection of
these members, in all the examples referred to, taking
the axis of the column as the line from which the measurements are made.

TABLE OF PROJECTIONS IN THE DORIC ORDER.

	Temple of Theseus at Athens.			Temple of Minerva at Athens.			Theatre of Marcellus at Rome.			Thermæ of Diocletian at Rome.		
	Projection from the axis of the Column.			Projection from the axis of the Column.			Projection from the axis of the Column.			Projection from the axis of the Column.		
	m.	p.	f.	m.	p.	f.	m.	p.	f.	m.	p.	f.
Entablature { Cornice	2	4		1	26		2	25		2	13	
Frieze		29			28			27			27	
Architrave	1	2			—			27		1	1	
Column { Capital	1	4		1	2		1	8		1	9	
Shaft	1	23		1	23		1	23		1	25	
Intercolumniation from axis to axis.	5	6	—	4	20	—	—			—		—

TABLE OF PROJECTIONS IN THE IONIC ORDER.

		Temple of Ilissus at Athens.			*Temple of Minerva Polias at Athens.*			*Temple of Fortuna Virilis at Rome.*			*Temple of Marcellus at Rome.*			*Colosseum at Rome.*		
		Projection from the axis of the Column.			Projection from the axis of the Column.			Projection from the axis of the Column.			Projection from the axis of the Column.			Projection from the axis of the Column.		
		m.	*p.*	*f.*	*m.*	*p.*	*f.*	*m.*	*p.*	*f.*	*m.*	*p.*	*f.*	*m.*	*p.*	*f*
Entablature	Cornice	2	11	—	2	7	½	2	26	¼	3	—	—	2	17	½
	Frieze		28	¾	1	—	—		28	—		29	—		27	—
	Architrave	1	2	½	1	7	—	1	2	¾	1	7	—	1	6	½
Column ...	Capital	1	1	—	1	—	—	1	7	¼	1	5	½	1	8	½
	Shaft		25	½		25	—		26	¼		25	¼		27	—
		1	—	—	1	—	—	1	—	—	1	—	—	1	—	—
	Base	1	11	½	1	15	—	1	13	—	1	10	½	1	10	—
Pedestal ..	Cornices		—			—		2	5	—	1	24	—		9	½
	Die		—			—		1	13	—	1	11	—	1	10	—
	Base		—			—		2	8	¾		—		1	20	—
Volute		1	15	¾	1	15	¼	1	11	¾	1	12	½	1	10	—
Intercolumniations from axis to axis		6	15	—	8	—	—	7	5	—		—			—	

TABLE OF PROJECTIONS IN THE CORINTHIAN ORDER.

		Choragic Monument of Lysicrates.			*Arch of Theseus at Athens.*			*Temple of Jupiter Olympus at Athens.*			*Temple of Jupiter Stator at Athens.*			*Temple of Jupiter Tonans at Rome.*			*Portico of the Pantheon at Rome.*		
		Projection from the axis of the Column.			Projection from the axis of the Column.			Projection from the axis of the Column.			Projection from the axis of the Column.			Projection from the axis of the Column.			Projection from the axis of the Column.		
		m.	*p.*	*f.*	*m.*	*p.*	*f.*	*m.*	*p.*	*f.*	*m.*	*p.*	*f.*	*m.*	*p.*	*f.*	*m.*	*p.*	*f.*
Entablature	Cornice	2	15	—	2	21	—	2	19	—	3	3	¾	2	13	—	2	16	—
	Frieze		29	¼	1	9	½	1	2	⅓		26	¾		26	—		25	½
	Architrave	1	3	½	1	8	¾	1	9	⅓	1	3	¾	1	4	⅓	1	3	½
Column	Capital	1	17	—	1	22	½	1	25	—	1	14	—	1	14	⅓	1	15	—
	Shaft		25	—		27	½		26	½		25	¾		26	—		26	—
		1	—	—	1	—	—	1	—	—	1	—	—	1	—	—	1	—	—
	Base	1	15	—	1	12	½	1	12	—	1	12	—	1	11	½	1	11	—
Pedestal	Cornice		—		1	25	½	1	13	—		—			—			—	
	Die		—		1	12	½	1	5	½		—			—			—	
	Base		—		1	19	¾	1	14	½		—			—			—	
Intercolumniation from axis to axis			6		8	—	—		—		5	5	—	5	3	½	6	10	—

English Architecture.

In closing our remarks upon architecture, it will be necessary for us briefly to refer to some of the most remarkable structures in England, and the best manner in which this can be done is to take a general view of the history of English architecture, It will not, however, be necessary that we should lead the reader through a long disquisition upon Druidical remains, and the character of the architecture introduced by the Romans when they became masters of the Island. At a very early period after the Romans left Britain an unprotected prey to the avaricious and warlike tribes that then roamed over the seas, and sought to aggrandize themselves by inroads upon weak and defenceless nations, the Saxons took possession of the country and maintained sovereignty over it, for a considerable period. But they were afterwards overcome by the Normans, who, from that time, became masters. By the Saxons, in all probability, a new style of architecture was introduced, and this was greatly improved upon by the Normans, who built large churches, and paid great attention to the progress of the art. This style continued to improve, until, in the time of the Henrys and Edwards, it arrived at maturity. In the reign of Elizabeth, classical literature revived in this country, and at that time the Italian architecture was introduced, and the beautiful style of the middle ages fell into disuse. In the succeeding ages arose many great men who, fascinated with the revival of classical literature, could not imagine any thing good that had not been done by the Greeks and Romans; and to them we are perhaps indebted for the term Gothic, the name which was given to the English style of architecture. When we consider the state of literature at the time in which Wren lived, we are not much surprised to find him stating that Gothic is a gross concameration of heavy, melancholy, and Monkish piles. But that any person professing an acquaintance with architecture, and offering to instruct others, should hold the same opinions, is, we confess, a phenomenon for which we cannot account. Can any man who has eyes, capable of conveying an impression, and a soul fit to enjoy the sensations of beauty and sublimity, visit the sacred piles of York, Lincoln, or Salisbury, and not feel that art has the mastery of him? If they convey an impression of gloom, it is to the dull sordid soul, never formed to enjoy the pleasures which poetry and art afford. We are sorry to find a modern author, whose writings we esteem, indulging his prejudices so far as to deny the value of this grand, impressive, and pure style of architecture. " The Gothic mode of construction," he says, "originated in a corrupted taste, and an ignorance of the original rules and sentiments; it is a sort of monster engendered by a chaos of ideas in the night of barbarity; a heterogeneous mixture of confused resemblances, obliterated traditions, and disparites of models. Far from being able to trace in it the first steps of a new and rising taste, it exhibits the impotency of one that is aged, lingering in darkness on the effaced traces of a model which has disappeared." If Gothic architecture, in its worst period, violated to one hundredth degree the principles of taste, as this author does the rules of good sense and good writing in the passage we have quoted, we would never again say one word in its praise. Such senseless bombast was never surely uttered by the lips or recorded by the pen of man, and it seems almost unnecessary that we should say one

x

word upon opinions so evidently formed by prejudice,
and stated in the spirit of a partisan. Who can tell
what the writer means by *original* rules and *sentiments?*
And who in his sober senses can imagine a monster en-
gendering chaos of ideas in a night of barbarity? Or
one that is aged, lingering in darkness *on the effaced
traces* of a model which has *disappeared?* If the
reader will only judge of the objections made to Gothic
architecture by the style and nonsense in which the
author, whom we have quoted here, indulges, he will
form as high an opinion of it as we have done.

The elements of Gothic architecture are spires, pin-
nacles, and pointed arches; and the cone, or pyramid,
seems to be the geometrical figure upon which it is
formed. In England it has been followed with more
success than in any other country, and we may anticipate that York Minster, Westminster Abbey, the Chapel
of Henry the Seventh, the Abbey Church of St. Albans,
and many other fine structures, will be visited for ages
to come, as the best models for the architectural talent
of Europe.

Since the time of Charles the Second, many fine
buildings have been erected in the British Isles, and
we may especially mention among these, Greenwich
Hospital, designed by Inigo Jones. A description of
any of the British works, whether in the English or
classical styles, cannot however be attempted in this
work; but if the sketch we have given should excite
the interest and enthusiasm of any reader, and induce
him to investigate for himself, it will have accomplished
the object for which it was written.

TERMS USED IN BUILDING.

A.

Abacus.—The upper member of the capital of a column, that on which the architrave rests. It has different forms in the several orders: in the Tuscan or Doric it is a square tablet; in the Ionic, its edges are moulded; in the Corinthian, its sides are concave, and frequently enriched with carving.

Abutment.—That part of a pier from which the arch springs.

Acanthus.—A plant, whose leaves are carved on the Corinthian and Composite capital. They are differently disposed, according to circumstances; and the leaves of the laurel and parsley are sometimes employed in their place.

Acroterium.—A pedestal on the angle or apex of a pediment, intended as a base for sculpture.

Altitude.—The perpendicular height of any thing in the direction of a plumb-line. The length of a body is measured on the body itself, and remains constant, its altitude varies according to its inclination to or from the perpendicular.

Alto Relievo.—A sculpture, the figures of which project boldly from the surface on which they are carved.

Amphiprostylos.—An order of Grecian temples, having columns at the back as well as the front.

Amphitheatre.—A double theatre, employed by the ancients for public amusements. The Coliseum, at Rome, built by Vespasian, is one of these.

Annulet.—A small square moulding, used to separate others; the fillet which separates the flutings of the column is sometimes known by this term.

Antæ.—Pilasters attached to a wall, receiving an entablature, and having bases and capitals differing according to the order employed, but always unlike those of the columns.

Antepagmenta.—A term in ancient architecture : the architraves round doors.

Apophyge.—That part of a column which connects the upper fillet of the base, and the under one of the capital with the cylindrical part of the shaft.

Arœostylos.—That style of building in which the columns are distant from one another from four to five diameters. Strictly speaking, the term should be limited to an intercolumniation of four diameters, which is only suited to the Tuscan order.

Arch.—Such an arrangement, in a concave form, of building materials, as enables them, supported by piers or abutments, to carry weights and resist strains.

Arch-buttress.—Sometimes called a flying buttress ; an arch springing from a buttress or pier, against a wall.

Architrave.—That part of the entablature which rests upon the capital of a column, and is beneath the frieze. It is supposed to represent the principal beam of a timber building.

Area.—This term is applied to superficies, whether of timber, stone, or other material, and is the superficial measurement; that is, the length multiplied into the breadth. The word area sometimes signifies an open space.

Arris—The line in which two surfaces meet each other.

Ashler.—Common freestone, as it comes from the quarry, generally about nine inches thick, but of different superficial dimensions.

Ashtering.—Quartering, to which laths are nailed.

Astragal.—A small moulding with a semicircular profile, sometimes plain and sometimes ornamented.

Attic Order.—A term used to denote the low pilasters which are placed over orders of columns or pilasters, and frequently employed in the decoration of an attic.

B.

Baluster.—A small pillar or pilaster supporting a rail.

Balustrade. A series of Balusters connected by a rail.

Band.—A square member. To distinguish the situation in which it is placed, or the order in which it is used, an adjective is frequently prefixed : thus a dentil or modillion band.

Base.—The lower division of a column. The Grecian Doric has no base, and the Tuscan has only a single torus on a plinth.

Bead.—A circular moulding, which lies level with the surface of the material in which it is formed. When the moulding projects, or several are joined, it is called reeding.

Beak.—A small fillet in the under edge of a projecting cornice, intended to prevent the rain from passing between the cornice and facia.

Beam.—A piece of timber in a building laid horizontally, and intended to support a weight, or to resist a strain.

Beam-filling.—The masonry, or brickwork, between beams or joists.

Bearer.—A vertical support.

Bearing.—The length between bearers, or walls; thus, if a beam rests upon a wall twenty feet apart, the bearing is said to be twenty feet.

Bed Mouldings.—Those mouldings between the corona and the frieze.

Bevil.—An instrument used by workmen for taking angles. In form it resembles a square, but the blade is moveable about a centre. When the two sides of any solid body have such an inclination to each other as to form an angle greater or less than a right angle, the body is said to be bevilled.

Bond.—A term used to signify the connexion between the parts of a piece of workmanship. In bricklaying and masonry, it is that connexion between bricks, or pieces of stone, which prevents one part of the building from separating itself from another.

Bond Timber.—Timber laid in walls to tie or bind them together.

Brace.—A piece of timber placed in an inclined position, and used in partitions or roofs, to strengthen the framing. When a brace is employed to support a rafter, it is called a strut.

Bressummer.—A beam, or iron tie, intended to carry an external wall, and itself supported by piers or posts.

Bricknoggin.—Brickwork between quartering.

Buttress.—A mass of stone or brickwork intended to support a wall, or to assist it in sustaining the strain that may be upon it,

x*

Buttresses in Gothic architecture, are used for ornament as well as strength.

C.

Cabling.—Cylindrical pieces filling up the lower part of the flutes of a column.

Camber.—To give a convexity to the upper surface of a beam.

Cantalivers.—Pieces of wood or stone beneath the eaves to support them, or mouldings above them.

Capital.—That part of a column or pilaster beneath the entablature; or, in other words, the uppermost member of a column or pilaster. Capitals are variously formed, according to the order: thus, we have the Tuscan, Doric, Ionic, Corinthian, and Composite capitals, and many others, that have been invented since the times of the Greeks and Romans.

Caryatides.—Figures of women, introduced, instead of columns, to support an entablature.

Casement.—Applied to a window which is hung upon hinges in place of lines and weights.

Casting.—The warping or shrinking of timber or woodwork, occasioned by an insufficient strength, or by an unequal exposure to the weather, and want of proper seasoning.

Cavetto.—A concave moulding, the quadrant of a circle.

Centering.—The framing upon which an arch is turned.

Clamping.—When one piece of wood is so fixed into the end of another as to prevent it from splitting or casting, it is said to be clamped. The pieces may be united with a mortice and tenon, or with a groove and tongue.

Collar Beam.—A beam framed between two principal rafters.

Console.—An ornament cut on the key-stone of an arch, sometimes in the form of a scroll, at other times to represent a human face.

Content.—The amount of any substance in rods, yards, feet, or inches, whether solid or superficial.

Coping.—The stone which covers the top of a wall or parapet.

Corbel.—A bracket, or piece of timber projecting from a wall: in Gothic architecture, usually carved with some grotesque figure.

Cornice.—The combination of mouldings which finishes or crowns an entablature. The term is also applied to the mouldings which finish the walls and ceiling of a room, hall, or passage, filling up the angle which they make.

Crown.—A term applied to the uppermost or highest part of an arch, that in which the key-stone is fixed.

Cyma.—A moulding with a waved or crooked profile, partly convex, partly concave. It is called by workmen an ogee. When the hollow part of the moulding is uppermost, it is called a cyma-recta ; when the convex part is above, a cyma-reversa.

D.

Dado.—That flat part of the base of a column between the plinth and the cornice. It is of a cubical form, and thence takes its name.

Dentils.—Square blocks introduced as ornaments into cornices of the Doric, Ionic, and Corinthian orders. A small circular piece is sometimes cut out, and at other times they are fluted.

Die.—A square cube.

Door Frame.—The case in which a door opens and shuts, consisting of two uprights and one horizontal piece, connected together by mortices and tenons.

Dormer.—A window made in the sloping part of a roof, or above the entablature.

Dovetailed.—When two pieces of wood are fastened together, by letting the pieces of one into apertures formed in the other, of a shape somewhat resembling a fan or dovetail, they are said to be dovetailed.

Drops.—Ornaments in the Doric entablature, resembling bells placed immediately under the triglyphs.

Dwarf Wall.—A wall that has a less height than that of the story in which it is used.

E.

Eaves.—The edge of a roof or slating which overhangs a wall, and is designed to carry off the water, without flowing down the wall.

Echinus.—A moulding, the profile of which is the quadrant of a circle turned outwards, or in some instances a conic section. It is said to resemble the shell of the chesnut.

Ellipse.—That curve, called by workmen, an oval.

Entablature.—That assemblage of mouldings, &c., which are supported by the column. It consists of three parts, the architrave, the frieze, and the cornice.

Entasis.—The swelling of a column.

Eustylos.—That intercolumniation, in which the columns are placed two diameters and a quarter from each other.

Eye.—A term sometimes used in architecture to denote a small window in a pediment. The middle of the Ionic volute, that is, the circle within which the different centres for drawing it are found, is known by the same name.

F.

Façade.—The face or front of a building; strictly speaking, the principal front.

Fascia.—A flat broad member, in architecture, but of small projection. It is used to denote the flat members into which the architrave is divided, and these are called fasciæ.

Feather-edged.—Boards or planks, thicker at one edge than the other.

Fillet.—A small square moulding, of slight projection. In carpentry, it means a piece of wood to which boards are nailed.

Flashings—Pieces of lead so let into the wall, as to lap over a gutter.

Flatting.—Painting, which has no gloss on its surface, being worked with turpentine. It is used for finishing.

Flutes.—Vertical channels, cut in the shafts of columns and pilasters; sometimes meeting each other at a sharp edge, and at other times having a fillet between them.

Flyers.—Stairs which rise without winding.

Flue.—The aperture of a chimney.

Footings.—The courses of brick or stone, at the foundation of a wall.

Frieze.—The flat member in an entablature, separating the architrave from the cornice.

Furring.—A means of restoring an irregular framing, by the addition of small pieces of wood nailed to the framing itself.

Fust.—The shaft of a column.

G.

Gable.—The upright triangular end of a building at the ends of a roof.

Girder—The largest piece of timber in a floor, that into which the joists are framed.

Groin.—The intersection of two arches.

Groove.—A rectangular channel cut in stone or timber; such as that which is cut in the stiles to receive the panel of a door.

Grounds.—Those pieces of wood embedded in the plastering of walls, to which skirting and other joiner's finishings are attached.

Guttæ.—See "*Drops.*"

Gutter.—A valley between the parts of a roof, or between the roof and parapet; intended to carry off the rain.

H.

Half Round.—A moulding in a semicircular form, projecting from the surface.

Headers.—Bricks laid with their short face in front.

Hips.—Those pieces of timber placed in an inclined position as the corners or angles of a roof.

I.

Jambs.—The side pieces of an opening in a wall, such as doorposts, and the uprights at the sides of window frames.

Impost.—The combination of mouldings which form the capital of a pier.

Insulated.—A term applied to a column which is unconnected with a wall; or to a building that stands detached from others.

Intercolumniation.—The space between two columns.

Intertie.—Small pieces of timber, placed horizontally between and framed into vertical pieces to tie them together.

Joggle-piece.—A post to receive struts.

Joists.—Those pieces of timber which are framed into a girder, bressummer, or otherwise, to support a ceiling or a floor.

K.

Key-stone.—That stone in the top or crown of an arch, which is in a perpendicular line with the centre.

King-post.—The centre-post of a trussed framing, intended to support the tie-beam and struts.

Knee.—A piece of timber bent to receive some weight or to relieve a strain.

L.

Lantern.—A frame in the dome, or cupola of a building, to give light. The term is applied to some kinds of fanlights, that is, the frame over a door to light a passage or corridor.

Lining.—That joiner's work which covers an interior surface.

Lintels.—The pieces of timber which lie horizontally over the jambs of windows and doors,

M.

Mantle.—The cross-piece which rests on the jamb of a chimney.

Metopa.—The interval between the triglyphs in the Doric order.

Minute.—The sixtieth part of the diameter of a column.

Modillion.—An ornament in the Ionic, Corinthian, and Composite orders. It is a sort of bracket, and should be placed under the corona.

Module.—The semidiameter of a column, and is divided into thirty minutes. It is the measure by which the architect determines the proportions between the parts of an order.

Mortise.—A method of joining two pieces of wood; a hole being made in one of such a size as to receive the tenon or projecting piece formed on the other.

Mosaic.—A term applied to pavements, and other work, when

formed of various materials of different shapes and colours, laid in a kind of stucco, so as to present some pattern or device. The ancients were very successful in the execution of Mosaic, and many fine specimens remain to this day.

Mullion.—Upright posts or bars, which divide the lights in a Gothic window.

N.

Naked.—This term is applied, in architecture, to a plain surface, or that which is unfinished ; as the naked walls, the naked flooring,—that is, uncovered. The word is sometimes applied to flat surfaces before the mouldings and other ornaments have been fixed.

Newel.—The centre round which the stairs wind in a circular staircase.

Nosings—The rounded and projecting edges of the treads of stairs.

O.

Obelisk.—A slender pyramid.

Ogee.—A moulding, consisting of a portion of two circles turned in contrary directions, so that it is partly concave and partly convex, and somewhat resembles the letter S.

Order.—An assemblage of parts, having certain proportions to one another. There are five orders of architecture, Tuscan, Doric, Ionic, Corinthian, and Composite, all of which were invented by the ancients, and are now employed by the moderns.

Oval.—A curve-line, the two diameters of which are of unequal length, and is allied in form to the ellipse. An ellipse is that figure which is produced by cutting a cone or cylinder in a direction oblique to its axis, and passing through its sides. An oval may be formed by joining different segments of circles, so that their meeting shall not be perceived, but form a continuous curve line. All ellipses are ovals, but all ovals are not ellipses ; for the term oval may be applied to all egg-shaped figures, those which are broader at one end than the other, as well as to those whose ends are equally curved.

Ovolo.—A convex projecting moulding, whose profile is the quadrant of a circle.

P.

Panel.—A compartment enclosed in a frame, into which it is framed or grooved.

Parapet.—A low wall generally about breast high, on the top of bridges or buildings.

Pargetting.—Rough plastering, commonly adopted for the interior surface of chimneys.

Pedestal.—That arrangement on which columns are sometimes placed.: it is divided into three parts, the cornice, the die, and the base.

Pediment.—A low triangular crowning ornament in the front of a building, and over doors and windows. Pediments are sometimes made in the form of a segment of a circle.

Pier.—A square, or other formed mass, used to strengthen or support a building; it sometimes signifies that mass of stone or brickwork between the arches of a bridge, and from which they spring, or against which they shut. But the term is usually employed to designate the solid part between the doors or windows of a building.

Pilaster.—A square pillar insulated, or engaged to the wall, and is usually enriched with a capital and base.

Piles.—Large timbers, usually shod with pointed iron caps, driven into the ground for the purpose of making a secure foundation.

Pillar.—An irregular, insulated column. It differs from a column in having no architectural proportion, being either too massive or too slender.

Pinnacle.—A small spire, used to ornament Gothic buildings.

Pitch of a roof.—The proportion obtained by dividing the span by the height : thus we speak of its being one-half, one-third, one-fourth.

Plinth.—The solid support of a column or pedestal.

Plumb-line.—An instrument to determine perpendiculars: it consists of a piece of lead attached to a string.

Porch.—The vestibule or entrance to a building.

Portico.—A kind of gallery or piazza, frequently erected in front of large buildings.

Posts.—Square timbers set on end; it is especially applied to those which support the corners of a building, and are then framed into the bressummer or cross-beam under the walls.

Pricking up.—The first coat of plaster worked upon laths.

Profile.—The outline; the contour of a part, or the parts compassing an order.

Pugging.—The stuff laid upon sound boarding to prevent the passage of sound from one story to another.

Puncheons.—Short pieces of timber employed to support a weight when the bearing is too distant.

Purlines.—Those pieces of timber which lie across the rafters to prevent them from sinking.

Putlogs.—Pieces of timber used in building a scaffold; they are those which lie at right angles to the line of a wall and rest on the scaffold poles or ledgers.

Pyramid.—A solid massive edifice, which rises from a square or rectangular base, and terminates in a point, called the vertex.

Q.

Quarter-round.—See Ovolo.

Quarters.—Pieces of timber used in an upright position for partitions. Quarters may be either single or double; the single are generally two inches thick, and four inches broad; the double, four inches square. The quarters are never placed at a greater distance than fourteen inches from each other.

Quirk.—A piece of ground taken out of a plot. The term is also applied to a particular form of moulding, one which has a sudden convexity.

Quoins.—The corners of a building; they are called rustic quoins when they project from the wall and have their edges chamfered off.

Y

R.

Rabbet or Rebate.—A groove or channel in the edge of a board.

Rafters.—Those timbers which form the inclined sides of a roof.

Raking.—Literally means inclining, and is applied to those mouldings which, instead of maintaining the horizontal line, are suddenly bent out of their course.

Rails.—Those pieces in framing which lie in a horizontal position are called rails; those which are perpendicular are called stiles: hence two rails and two stiles enclose a panel. The term is also applied to those pieces in fences or paling which go from post to post.

Relief.—The projection which a figure has from the ground on which it is carved.

Return.—That part of any work which falls away from the line of front.

Ridge.—The highest part of a roof, or the timber against which the rafters pitch.

Riser.—That board, in stairs, set on edge, under the tread or step of the stair.

Rustic.—This term is applied to those courses of stone work, the face of which is jagged or pecked so as to present a rough surface. That work, also, is called rustic, in which horizontal and vertical channels are cut in the joinings of the stones, so that when placed together, an angular channel is formed at each joint.

S.

Sash.—The frame-work which holds the squares of glass in a window.

Sash frame.—The frame which receives the sash.

Scantling.—The measure to which a material is to be or has been cut.

Scotia.—A semi-circular concave moulding, chiefly used between the tori in the base of a column.

Scribing.—Fitting wood-work to an irregular surface.

Scroll.—A carved curvilinear ornament, somewhat resembling in profile the turnings of a ram's horn.

Sill.—The horizontal piece of timber at the bottom of framing ; the term is chiefly applied to those pieces of timber or stone at the bottom of doors or windows.

Shaft.—The body of a column ; that part between the base and capital.

Shore.—A piece of timber placed in an oblique direction to support a building or wall.

Skirting.—The narrow boards placed round an apartment against the walls, and standing vertically on the floor.

Sleepers.—Pieces of timber placed on the ground to support the ground-joists, or other wood-work.

Soffit.—A term applied to a frame or panelling overhead, or to a lining, such as that which is fixed in the underside of the tops of windows.

Stiles.—The upright pieces in framing or panelling.

Strutts.—Pieces of timber which support the rafters.

Summer.—A large piece of timber supported by piers or posts : when it supports a wall, it is called a breast-summer, or bressumer.

T.

Tenon.—A piece of wood so formed as to be received into a hole in another piece, called a mortise.

Throat.—That hollow which terminates the upper end of the shaft of a column.

Tongue.—That projecting piece at the end of a board which is formed to be inserted into a groove.

Torus.—A moulding that has a convex semi-circular, or semi-elliptical profile.

Transom.—A piece that is framed across a double window-light.

Trellis.—An open framing, pieces crossing each other so as to form a diamond or lozenge-shaped openings.

Triglyphs.—Ornaments in the Doric frieze consisting of a square projection with two angular channels, the edges of each forming half a channel. They are placed immediately over the centre of a column : their width is generally one module.

Trimmers.—Pieces of timber framed at right angles to the joists for chimneys, and the well-holes of stairs.

Tympanum.—The space enclosed by the inclined and horizontal sides of a pediment.

V.

Valley.—The space between two inclined sides of a roof.

Vaults.—Underground buildings with arched ceilings, whether circular or elliptical.

Vertex.—The top or summit of a pointed body, as of a cone.

Volute.—The scroll in the capital of the Ionic order.

Voussoirs.—The stones which compose the face of an arch, having a somewhat wedge-shaped form.

W.

Wall-plates.—The timbers built up with a wall, to carry the joists.

Weather-boarding.—Feather-edge boards, fixed vertically, so as to lap over each other.

Well-hole.—The aperture left in floors to bring up the stairs.

THE END.

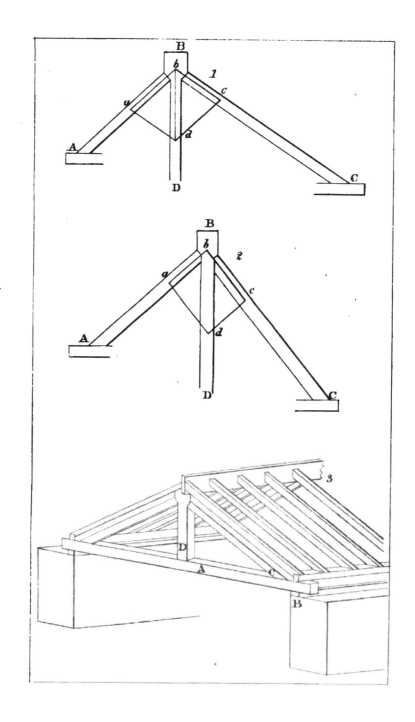

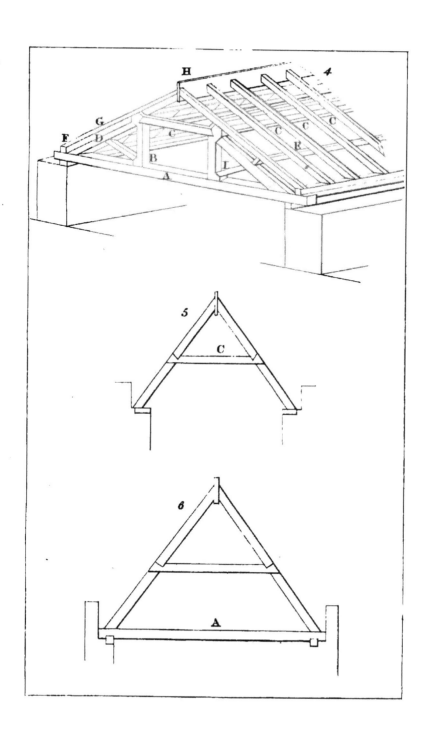

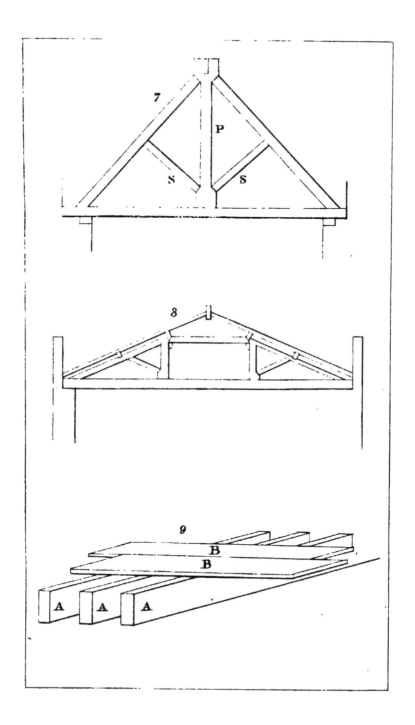

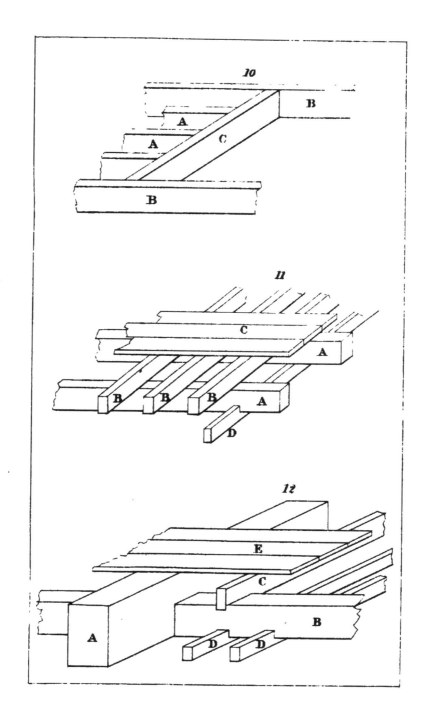

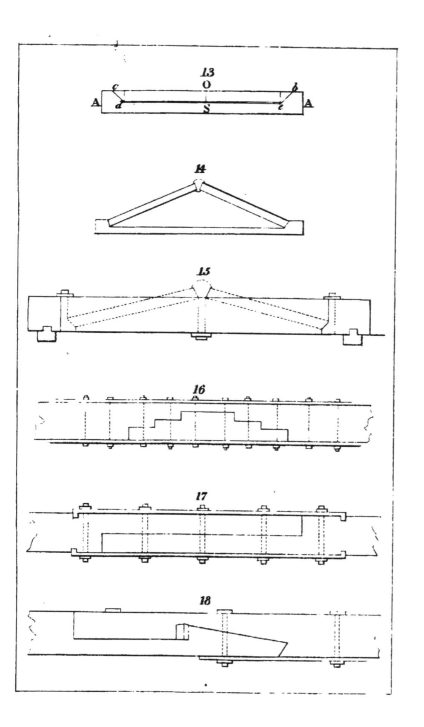

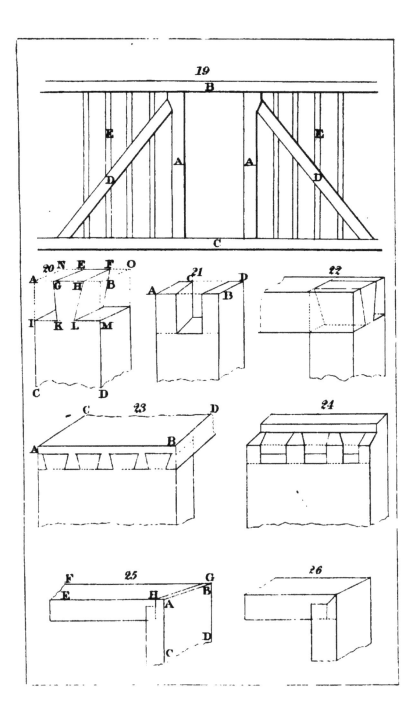

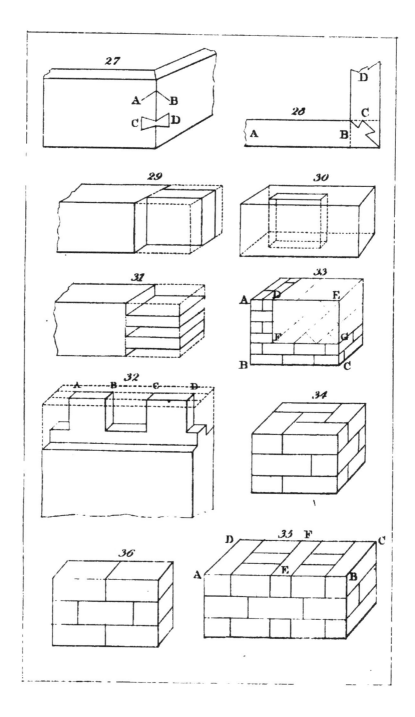

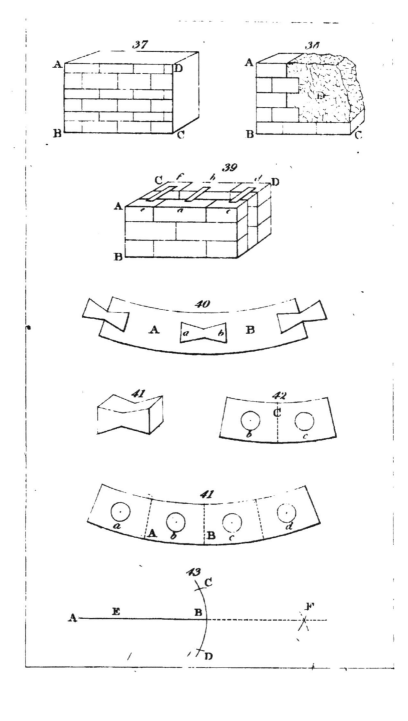

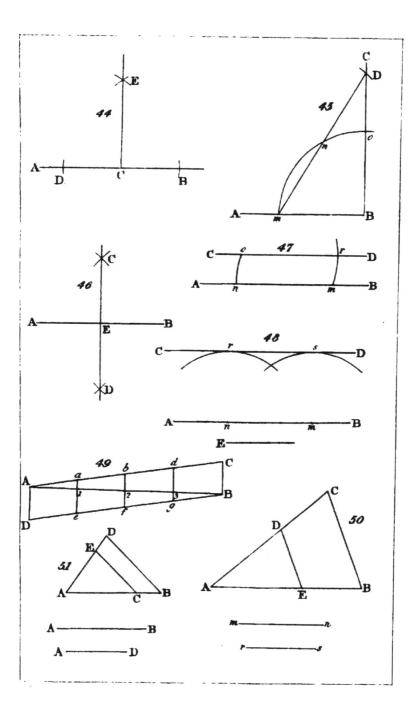

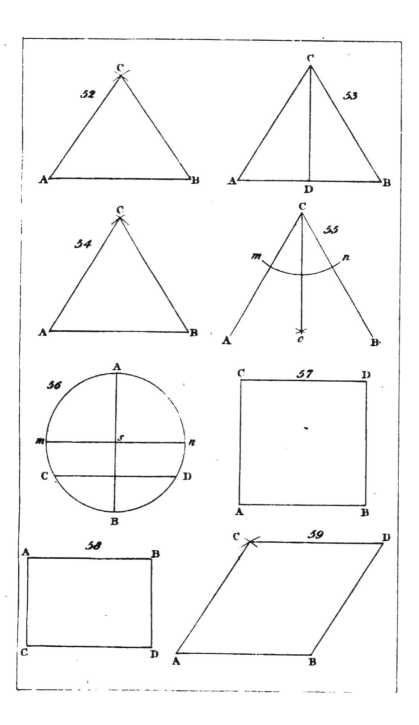

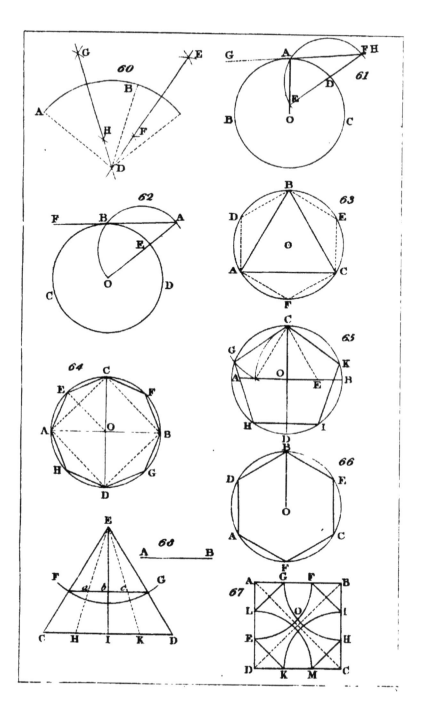

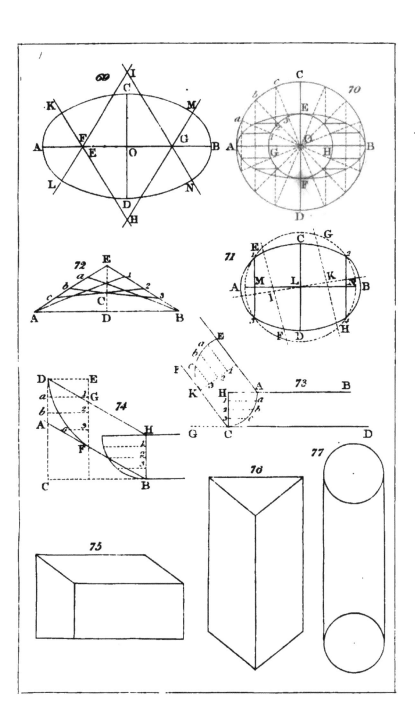